How NASA Learned to Fly in Space

An exciting account of the Gemini missions

by

David M. Harland

An Apogee Books Publication

Published by Apogee Books an imprint of Collector's Guide Publishing Inc., Box 62034, Burlington, Ontario, Canada, L7R 4K2, http://www.cgpublishing.com

Printed and bound in Canada

How NASA Learned to Fly in Space (First Edition) by David M. Harland

ISBN 1-894959-07-8 - ISSN 1496-6921

Contents

Author's Preface

For generations human beings stared at the Moon, dreaming of one day visiting it. Finally, with the orbiting of Sputnik in 1957, space travel began to look as if it would be feasible. In a remarkable speech in 1961, John F. Kennedy, the President of the United States of America, set his nation the goal of landing a man on the Moon before the decade was out. But how was this to be achieved?

The barely started Mercury missions were opening the door to manned space *craft*; the new program would have to pass through that door into the true realm of piloted space *flight*. It was decided to introduce an interim program between Mercury and Apollo, called Gemini, to develop the operational techniques necessary for spaceflight. This would 'write the book', yield a wealth of technology, a pool of flight controllers and a cadre of experienced astronauts for the Apollo program. With 10 manned Gemini missions in 1965-1966, the pace was hectic, but by the time it was over, orbital rendezvous was no longer the concern that it had been in 1962 when NASA had committed Apollo to rendezvous in lunar orbit. So the foundation of Gemini's legacy was the security of knowing that Apollo was feasible. As Dr Robert Gilruth, director of the Manned Spacecraft Center in Houston, observed afterwards: "In order to go to the Moon, we had to learn how to operate in space. We had to learn how to manoeuvre with precision to rendezvous and to dock; to work outside in the hard vacuum of space; to endure long-duration in the weightless environment; and to learn how to make precise landings from orbital flight – that is where the Gemini Program came in."

This book *isn't* an official history, nor does it delve into the finances of the Program, nor does it trace the paper trail through the decision-making. It *is* the story of how NASA learned to fly in space. It is set in a time when the agency was young and lean, and had an explicit mandate of staggering audacity set against a tight deadline; in a time when the agency readily accepted risk, and made momentous decisions 'on the run' and successfully followed through; in a time when a rendezvous was a major objective of a mission, when opening the hatch and venturing outside was a serious challenge. Apollo claimed the glory, but it was Gemini which 'stretched the envelope' of spaceflight to make going to the Moon feasible. This, therefore, is the story of these missions. As it is a story of exciting times, I have drawn on the transcripts to recreate the sense of drama. Quotations have been edited for clarity, for brevity, and to eliminate the intermingling that is characteristic of spontaneous conversation. Nevertheless, I have preserved the sense of the moment. I have also attempted to explain orbital rendezvous in order to enable the reader to share in the astronauts' delight when things went to plan and their frustration when – as happened all too often – they did not.

David M Harland
Kelvinbridge, Glasgow
April 2004

Acknowledgements

This is one of those books which has been years in the making (in fits and starts) and a great many people have lent their assistance along the way. I would therefore wish to thank Roger Launius, Stephen Garber and Colin Fries of NASA's History Office in Washington D.C. for the hardcopy transcripts of the Gemini missions; Glen Swanson at the Johnson Space Center in Houston for supplementary archive documents; Mike Gentry of JSC's Media Services for arranging access to the image archive; Mike Smithwick for copying his Gemini 5 flight plan, complete with the crew's doodles and rhymes; Ed Hengeveld for scanning in pictures, and for allowing me to reproduce one of his paintings; Chris Gamble, John Pfannerstill, Roland Speth and Derek Henderson for providing additional material; Reg Turnill, formerly of the BBC, for supplying a most useful document from his archive; Frank O'Brien for assistance with some obscure documents; Jerry Bostick, Gemini Flight Dynamics Officer, for explaining some of the mysterious acronyms of rendezvous procedure; Dave Scott for his technical assistance with a multitude of arcane points; Sy Liebergot for help in tracking down the players; Eric Jones, Jim DeRuvo and Julian Bordas for their encouragement early on; Rich Orloff; Kipp Teague; John Catchpole for suggesting the title; Ken Glover, David Woods, Dick Glueck, Harald Kucharek, Philip Baker, Keith Wilson and Markus Mehring for reviewing early drafts of the text; Mike Hanlon for reviewing the final draft; Mark Gray of SpacecraftFilms.com for making high-quality film transfers onto DVD; and of course Rob Godwin of Apogee Books for making the book a reality.

To
Neil Armstrong and Dave Scott
who were nearly lost.

Chapter 1

Space Race

America Gears Up

The New Frontier

On 4 October 1957, to mark the International Geophysical Year, the Soviet Union placed the world's first artificial satellite into orbit, naming it 'Sputnik' (for 'fellow traveller') and the 'beeps' of its radio signal awed the populations of the countries over which it passed. After a spectacular 'flopnik', the United States sent up its own Explorer on 31 January 1958, which, in detecting the presence of charged particles trapped in the Earth's magnetic field, made the first significant discovery of the Space Age.

A wide range of options

Six weeks later, Maxime Faget at the Langley Research Laboratory in Virginia started to develop the specifications for a "manned satellite". Contrary to the prevailing belief that a spacecraft would be a derivative of the ever-higher ever-faster trend in aircraft, he decided to design a "capsule" that would be launched vertically on a rocket. In space, its orientation (its 'attitude') would be controlled by thrusters firing cold gas. On leaving orbit, it would make a ballistic re-entry and land by parachute. The main prerequisites were a rocket with sufficient thrust to achieve orbital speed, and the thermal shield to protect the capsule during re-entry. Powerful rockets were already under development. It was the thermal problem that posed the technological challenge. Faget decided on a conical capsule that would re-enter the atmosphere 'blunt end forward'. As it would create no aerodynamic lift, it would follow a ballistic course. The blunt shape would compress the air and generate a shock wave that would prevent most of the heat from impinging on the capsule, and the base was to be protected by an ablative shield which would prevent heat from penetrating the craft by flaking off into the slipstream. This concept was presented to an aerodynamics conference at the Ames Aeronautical Laboratory on 18 March in the paper entitled *Preliminary Studies Of Manned Satellites - Wingless Configuration, Non-Lifting*.

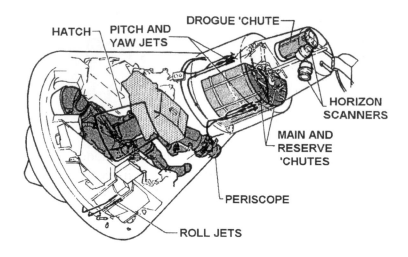

HATCH   PITCH AND   DROGUE 'CHUTE
        YAW JETS

                                        HORIZON
                                        SCANNERS

                                        MAIN AND
                                        RESERVE
                                        'CHUTES

                                PERISCOPE

                ROLL JETS

The internal layout of the Mercury spacecraft.

On 1 October, the National Aeronautics and Space Administration (NASA) was set up as a civilian agency to manage America's space projects under the leadership of Keith Glennan. A week later, it initiated Project Mercury "to achieve, at the earliest practicable date, orbital flight and successful recovery of a manned satellite." Unsaid was that this was to be achieved ahead of any Soviet effort to send a man into space. To develop an appropriate spacecraft the agency formed a Space Task Group at Langley under the leadership of Robert Gilruth, and in early November this invited concept proposals from industry for a spacecraft, and set out to recruit 'astronauts' to fly it. In December, eleven companies submitted bids for the spacecraft and on 12 January 1959 the contract was awarded to the McDonnell Aircraft Company in St Louis, Missouri. On 9 April NASA announced the names of the seven men, all of whom were military test pilots (three from the Air Force, three from the Navy and one from the Marine Corps) that it had selected as astronauts.

On 15 December 1958, Glennan received a presentation from Wernher von Braun of the Army's Ballistic Missile Agency in Huntsville, Alabama, where the Redstone used to put the Explorer satellite into orbit had been built. The Redstone had a thrust of 75,000 pounds. The Atlas missile being developed by the Air Force had a thrust of 360,000 pounds. Von Braun likened a rocket with a cluster of engines to multi-engined aircraft, and explained that he was working on a more powerful engine which, if clustered, would be able to yield a thrust of 1.5 million pounds. Furthermore, the previous year the Air Force had ordered the development of a *single* engine providing this much thrust, and a rocket with a cluster of such engines would be able to launch a space station or dispatch an expedition to the Moon.

A Research Steering Committee on Manned Space Flight was established in April 1959 under the chairmanship of Harry Goett at Ames to consider possible post-Mercury man-in-space objectives, which included:

* launching and operating a small orbital laboratory;
* assembling a large permanent space station;
* flying circumlunar and lunar orbital missions;
* making a lunar landing.

Having laid the groundwork on a heavy launch vehicle, and strategies for both assembling space stations and mounting flights to the Moon, von Braun's views strongly influenced the committee. The single most important factor in designing a mission to land on the Moon was called the 'mission mode'. The simplest strategy, known as Direct Ascent, was portrayed in the well-received 1950 science fiction movie *Destination Moon*, but von Braun calculated that it would require a rocket with a cluster of at least ten of the most powerful engines then under development. A more sensible solution, he insisted, would be to assemble the spacecraft from specialised components launched on a succession of smaller rockets, prior to departing for the Moon.[1] That summer, two groups at Langley independently investigated orbital rendezvous. John Eggleton studied the 'mechanics' of the task whilst John Bird considered trajectories and 'launch window' constraints. Looking to the longer term, John Houbolt set up a committee to investigate how rendezvous would influence mission strategies. One early outcome was that Langley urged the Goett Committee to make assembling a small orbital laboratory an essential intermediate objective, in order to be able to develop practical experience of rendezvous prior to mounting a lunar mission. In addition, in August Kurt Strass chaired a panel which mapped out a series of graded steps that would lead towards a lunar mission. It argued for a series of intermediate goals "to focus attention on problems to be solved, and thus serve to guide new technological developments". For example, once the Mercury spacecraft had been proven, it might be enhanced to spend longer periods in space, enlarged to accommodate two astronauts and given an engine to enable it to change its orbit. There was no shortage of ideas for future projects, but until they were given engineering rigour they were simply that – ideas. Whilst enhancing Mercury offered the benefit of building on what everyone hoped would soon be a proven vehicle, the risk was that the intrinsic limitations of the design might actually stifle progress in the longer term. Accordingly, the Space Task Group resolved on 2 November to develop a sophisticated three-man spacecraft that would be suitable for a variety of missions through the decade. Abe Silverstein, Director of NASA's Office of Space Flight Programs, suggested that this new spacecraft be named 'Apollo', and the name caught on.

The *Ten Year Plan* issued by the Goett Committee in December 1959 proposed working towards a "lunar reconnaissance" – a mission in lunar orbit. Lunar surface exploration was to be a "major goal for the ensuing decade". As 1960 dawned, therefore, NASA proposed as "a logical intermediate step" towards the "ultimate goal" of a lunar landing, the development the Apollo spacecraft that would see

[1] Whilst von Braun referred to this preparatory activity as 'orbital operations', it was soon renamed Earth Orbit Rendezvous because it would exploit space rendezvous.

astronauts flying in lunar orbit by the end of the decade. All of this project definition was being done without any official commitment to such a mission. Unfortunately, when the agency included start-up funding for Apollo in its Fiscal Year 1961 budget submission to President Dwight Eisenhower in early 1960, he refused. In establishing NASA and authorising Project Mercury, he reckoned that he had made the *least necessary* response to the humiliation of Sputnik. But to NASA, Mercury was the first step towards a glorious future! To legitimise what was being done to date on an *ad hoc* basis, on 3 January 1961 Glennan expanded the Space Task Group's mandate with planning "in the general area of manned space flight". Two days later Robert Seamans, the newly appointed Associate Administrator, asked George Low, in charge of Manned Space Flight, to write a report that drew together the options for a lunar landing. The studies at Langley and Huntsville had been conducted more or less independently. Low reasoned that since orbital operations would be necessary in the long term, rendezvous methods "must be developed" irrespective of whether a lunar landing was pursued. A few months later, in drawing up its Fiscal Year 1962 budget proposal, NASA factored in Low's recommendation for making an early start in developing rendezvous as a generic skill on the basis that doing so would open options.

On 20 January, Robert Gilruth convened a Space Task Group panel to discuss immediate post-Mercury options, focusing on hardware requirements of the spacecraft and the launch vehicle. Faget outlined two broad classes of mission, one to demonstrate rendezvous and the other to pursue extended duration in orbit. The publicly stated goal of Project Mercury was a three-orbit mission, but it was evident that with supplemental consumables it would be able to remain in space for a day or so. However, adding either an engine for orbital manoeuvring or the stores to sustain a week-long mission would push the spacecraft over the mass that the Atlas could place in orbit, in which case it would be necessary either to add an upper stage or to switch to a more powerful rocket. Faget was also concerned that the task of rendezvous in space might prove "too hazardous for a one-man operation", so he proposed making it into a two-man spacecraft, in which case a more powerful rocket would certainly be necessary. And if the Mercury spacecraft was to be adapted for more sophisticated missions, he had a list of "refinements" that would make it easier to assemble, check out, and service. On 1 February, Gilruth told James Chamberlin, Mercury's Engineering Director, to work with McDonnell to develop a detailed specification for an enhanced spacecraft. Although Chamberlin deliberately set out "to make a better mechanical design" without changing the overall form or the function of the vehicle, he soon came to the conclusion that it would have to be completely redesigned. He reported to Gilruth on 17 March. Prior to joining NASA in April 1959, Chamberlin had designed fire-control systems for fighters, systems which required frequent maintenance. The flaw with Mercury, he said, was that its electrical systems were stacked up inside the capsule in the manner of a multi-tiered cake, whereas they should be directly accessible from outside, as on an aircraft. He proposed stripping the spacecraft's systems down and repackaging them so that the spacecraft would not only be simpler to assemble but also easier to service on the launch pad. Like Faget, he wanted to transform an engineering test vehicle into an operational spacecraft. Gilruth accepted that Chamberlin had a case. On 12 April, McDonnell submitted a proposal for a detailed engineering study of such a redesign, which was approved two days later. Furthermore, the company was authorised to purchase long-lead-time items, so that, if NASA decided to pursue the idea, half a dozen of the 'new' spacecraft could be built without inordinate delay.

Kennedy's challenge

John F Kennedy took over as president in January 1961 after a campaign in which he had condemned Eisenhower for having allowed America to fall behind the Soviet Union in terms of bombers and strategic ballistic missiles; in fact, America was not trailing but this could not be revealed for national security reasons. As Kennedy had said, "The first man-made satellite was named 'Sputnik'. The first living creature in space was Laika. The first rocket to the Moon carried a red flag. The first photograph of the far side of the Moon was made with a Soviet camera. If a man orbits the Earth this year, his name will be Ivan!" Kennedy saw space as a new arena in the Cold War which would enable America to demonstrate its superiority without involving combat. As such, it was vital to national security that NASA seize the lead in space. Kennedy's rival, Nikita Khrushchev clearly realised the significance of *being first*. Eisenhower, Kennedy had argued, had failed to rise to this challenge. Kennedy argued that the first man in space must be an American, but on 12 April 1961, the first man to ride a rocket into orbit was called Yuri Gagarin.

On 12 April 1961 Yuri Gagarin became the first man to orbit the Earth.

The next day Kennedy asked Theodore Sorensen, a friend and advisor, to find "at what point we can overtake the Russians." Sorensen called a meeting with James Webb, who had been appointed in February to succeed Glennan, Hugh Dryden, who Webb had made his deputy, Jerome Weisner, who was Kennedy's science advisor, and David Bell, Kennedy's budget director. As Kennedy had rebuffed its Apollo lunar ambitions a month earlier, NASA pitched a form of the visionary plan that von Braun had published a decade earlier in the popular magazine *Collier's*, which envisaged assembling a space station as a 'jumping off' point for expeditionary missions to the Moon and the planets. However, if the Soviets were on the same plan they would likely remain in the lead for some considerable time. Kennedy wanted to minimise this period, either by accelerating or by 'short circuiting' the plan. Space was a race. The challenge that he set would have to be one that could be met decisively. "If somebody can just tell me how to catch up," he implored when he was briefed by Sorensen's working group. "There is nothing more important," he added after a moment's consideration. As the meeting broke up, Sorensen remained behind to discuss the options, and when he emerged he announced, "We're going to the Moon!"

On 5 May 1961 Al Shepard rode a Mercury spacecraft named 'Freedom 7' fired down the Eastern Test Range by a Redstone missile.

The ramifications of this decision were worked out over the next several weeks. Kennedy waited until Al Shepard restored some honour by making the first Mercury flight on 5 May – although his suborbital arc was hardly a match for Gagarin's orbit – and on 25 May he made a speech to a joint session of Congress on the subject of 'Urgent National Needs'.

"If we are to win the battle that is now going on around the world between freedom and tyranny..." Kennedy began, "it is time to take longer strides, time for a great new American enterprise, time for this nation to take a clearly leading role in space achievement which, in many ways, may hold the key to our future on Earth." But this was merely to set the mood. He followed up with an audacious punchline. "I believe this nation should commit itself, to achieving the goal, before this decade is out, of landing a man on the Moon and returning him, safely, to the Earth." He was challenging Khrushchev to a race to the Moon! But he was under no delusion that it would be straightforward. "No single space project in this period will be more impressive to Mankind. Or more important for the long-range exploration of space. And none will be so difficult or expensive to accomplish." To ensure that everyone understood that it was a matter of national honour, he added, "In a very real sense, it will not be one man going to the Moon... it will be an entire nation."

John F Kennedy speaking to a joint session of Congress on 25 May 1961: "I believe this nation should commit itself to achieving the goal, before this decade is out, of landing a man on the Moon and returning him, safely, to the Earth."

On being alerted to Kennedy's decision on 2 May, Seamans had asked William Fleming, Acting Assistant Administrator for Programs, to "detail a feasible and complete approach" to the accomplishment of an early lunar landing. Fleming drew his committee primarily from Headquarters and the Space Task Group. Despite the emphasis on an "early" mission, the committee opted for Direct Ascent as it thought that this would be "the simplest possible approach". It ruled out orbital operations because rendezvous was too great an unknown, and therefore an unjustifiable risk. On the day of Kennedy's historic speech, Seamans asked Don Ostrander, the Director of Launch Vehicle Programs, to form a committee "to assess a wider variety" of possible modes. Ostrander appointed the Lewis Research Center's Bruce Lundin to chair this committee. In contrast to Fleming, Lundin drew his membership from the various field centres and rejected Direct Ascent as impracticable in the timescale and focused instead on Earth Orbit Rendezvous. On 18 June, after receiving both reports, Seamans decided that he needed to even out the level of detail in the studies. Ostrander liaised with von Braun (who favoured Earth Orbit Rendezvous) to develop an overall mission plan. Ostrander's deputy, Donald Heaton, chaired a committee tasked with establishing the program requirements for a range of rendezvous-based modes. Heaton drew his membership from both Headquarters and the field centres in order to present a NASA-wide viewpoint, and reported in late August that "rendezvous offered the earliest possibility for a successful manned lunar landing." This was the opposite of Fleming's conclusion. The matter of the Apollo mission mode rested on two imponderables: (1) the engineering challenge of developing a superbooster and (2) the viability of orbital operations. Put simply, if rendezvous proved to be impracticable there would be no alternative to the development of a superbooster, but if the rendezvous difficulties had been overstated then the technologically challenging development of such a launch vehicle would be unnecessary. It was therefore crucial – as George Low had recommended – to find out as soon as possible whether rendezvous was viable.

Gemini is born

Meanwhile in St Louis, Chamberlin had been methodically working out how "to increase component and system accessibility" in the Mercury spacecraft, and when he reported to the Capsule Review Board on 9 June, Gilruth, Low and Faget were astonished by the scope of the changes. In "dissociating" what he saw as overly-integrated systems Chamberlin wanted to repackage – or modularise – almost every system and place it outside rather than inside the pressure shell. Furthermore, in a reversal of the original design requirement that the spacecraft be capable of flying automatically, Chamberlin wanted to delete the automatic sequencer that had proven so difficult to certify – "the root of all evil", he called it – and rely instead on the astronaut to control the vehicle. Only the external shell of the spacecraft had been sacrosanct, because he was not permitted to change the aerodynamics. Chamberlin had taken a particular dislike to the launch escape system. The Atlas burned kerosene with oxygen. This was such a volatile mixture that a tower with an escape rocket had been placed on the nose of the Mercury spacecraft to pull it clear of the fireball of an exploding booster. Quite apart from the fact that this represented dead weight, its sequencer was a nightmare to check out. In Chamberlin's view, if the spacecraft's functionality was to be upgraded, items like the automatic sequencers would have to be deleted. The constraint was the capacity of the Atlas rocket. The decision to use the Atlas had been straightforward, because it had been the most powerful missile available at the time. On 8 May, however, Albert Hall, the General Manager of the Martin Company's Baltimore Division had briefed Silverstein on its new Titan missile, which was considerably more powerful than the Atlas.

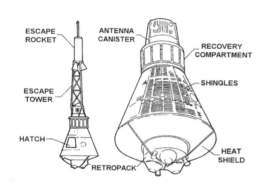

The Mercury spacecraft showing the launch escape system that would pull the capsule away from the launch vehicle in an emergency, and the retropack to de-orbit it at the conclusion of the mission.

As the Atlas intercontinental ballistic missile was being developed, the Air Force issued a contract to the Martin Company to design the Titan using a different design strategy, just in case the Atlas proved unsatisfactory. In the brief interval since the Atlas had been designed, the engineering database on rocketry had greatly improved. It was therefore decided to relax the constraint that had led to the Atlas being a 'one-and-a-half-stage' vehicle, in which all of the engines were ignited prior to lift off. Although air-starting a big liquid-propellant rocket had yet to be demonstrated, the Air Force eagerly accepted the tandem-staged configuration. In other respects, however, the design of the Titan was deliberately *less* innovative than the Atlas so as to minimise the technical risk – for example, instead of using the self-supporting pressurised tankage of the Atlas, a cylindrical aluminium airframe was to enclose the tanks, each of which would have a rigid internal structure. However, it was decided to use the same propellants, namely refined kerosene and liquid oxygen. The twin-chambered Aerojet LR87 engine of the first stage yielded 300,000 pounds-thrust and the single-chambered LR91 of the second stage produced 80,000 pounds-thrust. Instead of the small vernier engines used by the Atlas for trajectory control, it was decided to gimbal the engines for thrust-vectoring, but the second stage had verniers to refine the trajectory following the shut down of its main engine. The development of the Titan was rapid, and in early 1959 the first stage was tested using a dummy second stage, with ten full-range test flights following in 1960. As the Titan was a much more advanced design than the Atlas it had been expected to serve alongside the Atlas and then supersede it, but progress in developing large solid-propellant rockets meant that a new generation of intercontinental ballistic missiles became available much earlier than predicted, and rendered the Titan obsolete almost as soon as it was deployed operationally. In 1963, therefore, the Air Force decided to phase it out. Unlike the Atlas, the Titan was not converted for use as a launch vehicle, it was simply scrapped.

When the Titan had entered flight test, the Air Force ordered its successor. The Titan II was outwardly similar, but the engines burned aerozene (essentially unsymmetrical dimethyl hydrazine) with nitrogen tetroxide, and had been upgraded (the LR87 now produced 430,000 pounds-thrust and the LR91 produced 100,000 pounds-thrust). The operational advantage of these propellants was that they could be stored in the tanks (the cryogenic oxidiser of the original had to be loaded for launch) and it could be on its way within 60 seconds of the command being issued. As these propellants were hypergolic (meaning that they ignited on coming into contact) the engines did not require an igniter. Reliability was further enhanced by technical advances that facilitated a reduction in the number of relays, umbilicals, valves and regulators. Again, development was rapid, and the missile flew its 8,800 nautical mile target range by 1962. With a fully independent inertial guidance system, this fast-response missile could implement the massive retaliatory strike of the doctrine of 'mutually-assured destruction'. Furthermore, it incorporated a computer that permitted specification of one of three predefined targets for operational flexibility. Being much larger than its predecessor, it could lift a much heavier payload and was loaded with a 9 MT nuclear warhead – the most powerful yield of any American missile warhead before or since.

Impressed by the Titan II's performance, Silverstein asked Gilruth to consider man-rating it. Chamberlin seized on it as a means of escaping from the Mercury spacecraft's mass-limit. It was not just the raw performance that attracted him. The company had used its experience to *productionise* the missile with a simple rugged design that would make it straightforward to erect and check. But best of all, although the hypergolic propellants burned on coming into contact, they were less explosive, which meant that in a catastrophic malfunction there would be time for an ejection seat to carry the occupant clear. In Chamberlin's view, the missile was perfect for an operational spacecraft. It would be possible to dispense with the heavy launch escape system and its infernal sequencer. Furthermore, since the ejection seat would require a large hatch to be fitted, this could be designed to swing on a hinge instead of being bolted on, which would not only simplify the preparations for launch but also offer the prospect of the astronaut opening the hatch in space to venture outside. The Capsule Review Board met again on 12 June and decided that in the long term, if the spacecraft was to be made *operational*, the investment in an extensive redesign would pay off. However, without the funding for such an upgrade the Board restricted Chamberlin to developing the "minimum modifications" required to facilitate a 24-hour flight, which was something that seemed likely to be called for as soon as the first orbital mission was accomplished. Nevertheless, in July the Space Task Group started to explore the implications of Chamberlin's full proposal. On 7 July, Walter Burke of McDonnell gave a briefing in which he outlined three options. The first (which was what the company had been authorised to do) was the "minimum modification" for a 24-hour mission. The second option was to "reconfigure" the spacecraft as Chamberlin had outlined. As a final enhancement, Burke suggested adding a second seat. In fact, Faget had raised this possibility at the Capsule Review Board on 9 June, arguing that if they were going to overhaul the spacecraft to the extent proposed by Chamberlin then they might as well make it a two-man vehicle. He had already suggested that if rendezvous proved to be as tricky as some feared, it would be prudent to include a second pilot to share the load. And if the hinged hatch was to facilitate external activity, then the second pilot would be able to look after the spacecraft and to render assistance as necessary. There was considerable merit to modularising the spacecraft and upgrading its capabilities, and switching to the Titan II to accommodate the greater mass, but even after Kennedy instructed NASA to go to the Moon there was no *requirement* to do so. Gilruth and Silverstein went to St Louis on 27 July to review progress on the "minimum modifications". Whilst there, they were shown a wooden mock-up of the two-man spacecraft that Burke had outlined. It had large hinged hatches, and a panel between the seats on which was mounted an integrated hand-controller for the orbital manoeuvring system. Although not much larger than Mercury, it had a spectacularly different 'look and feel'. Wally Schirra, one of the astronauts, was impressed. The next day, Silverstein told the company to concentrate exclusively on the two-man variant, because it was simply too good a spacecraft to pass up. He could not have known just how soon this decision would be vindicated.

John C Houbolt of the Langley Research Center, who pushed for Apollo to utilise a 'mission mode' that involved Lunar Orbit Rendezvous.

On realising that the superbooster for Direct Ascent was unlikely to be developed in time to meet Kennedy's deadline of the end of the decade, NASA acknowledged that some form of orbital operations would be necessary. By now, however, there was a new option. In 1960 William Michael, a member of Houbolt's team that studied how rendezvous would influence mission strategies, had suggested Lunar Orbit Rendezvous. Instead of heading straight for the lunar surface, the spacecraft would enter lunar orbit and separate from its primary propulsive stage, make the landing on its own, lift off again and retrieve its propulsive stage to leave orbit to head for home. This traded off the propellant used to land the entire spacecraft as against the propellant required to enter orbit, prior to sending down only the section of the ship that *had* to land. Given that rendezvous had yet to be demonstrated, relying upon doing so around the Moon seemed, to some, to be inviting trouble because a crew stranded in lunar orbit would be doomed. As Houbolt worked through the implications he found that Lunar Orbit Rendezvous offered "a chain-reaction simplification," that gave "a very substantial saving in Earth-boost requirements". In fact, it enabled a lunar landing mission to be mounted by a single rocket of modest power, which eliminated the need to assemble the spacecraft in low orbit. Whilst the Goett Committee had defined the next major goals as circumlunar and lunar orbital missions, which could be achieved by a modest launch vehicle and Earth Orbit Rendezvous, Kennedy's commitment to a lunar landing before the decade was out made the choice of mission mode for Apollo of vital importance. A demonstration that rendezvous and docking was feasible would be a considerable comfort. It would be years before the first Apollo spacecraft rolled off the production line. The new sense of urgency resulted in the reversal of the decision to wait for Apollo before addressing rendezvous – it would have to be done by upgrading Mercury. This risk-mitigation exercise was approved on 7 December. As the two-man spacecraft was to be a modified Mercury (dubbed Mercury Mark II) the contract with McDonnell was signed on 22 December. At the behest of Alex Nagy of NASA HQ, on 3 January 1962 this

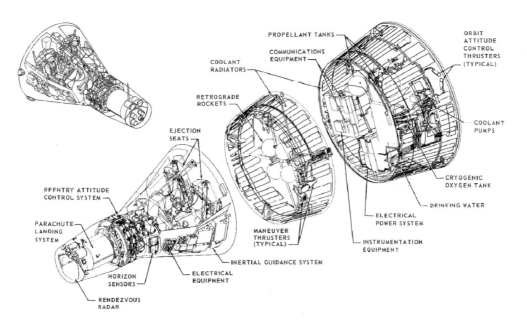

The three sections of the Gemini spacecraft: the re-entry module in which the crew rode and the retro and equipment sections of the adapter module.

program was named Gemini. Although an *interim* program, it would have a substantial budget because in addition to revising the Mercury spacecraft, the Titan II missile would have to be man-rated to carry it and a propulsive stage adapted to serve as a rendezvous and docking target. Given that an American had yet to fly in orbit, this was a very ambitious commitment! After much debate, the Manned Space Flight Management Council opted for Lunar Orbit Rendezvous for Apollo on 22 June, and announced this decision on 11 July. By then, however, planning was well underway.

Whereas when 1961 began, NASA had been uncertain as to whether there was a future for man-in-space beyond Mercury, by the end of the year it had two follow-on programs, one to facilitate the other. Kennedy's strategic vision had transformed the "ultimate goal" of a lunar landing into the primary objective.

With the commitment to Gemini made, Chamberlin turned to drawing up the preliminary project development plan.

In addition to incorporating the 'II' of 'Mercury Mark II', the Program patch represented the two astronauts by Castor and Pollux, the brightest stars of the constellation of Gemini - the twins of mythology.

An early idea by McDonnell for enhancing the Mercury spacecraft exploited the fact that the capsule was mated to the Atlas by a conical adapter in which an external module with a propulsion system and consumables for a long flight could be accommodated. Although the Atlas could not lift such a heavy package, the Titan II could. The new spacecraft comprised a re-entry module and an 'adapter' that mated with the 10-foot-diameter launch vehicle. In fact, this adapter was a composite. The equipment section had the consumables and thrusters for attitude control and orbital manoeuvring. It would be jettisoned prior to the de-orbit burn by the solid rockets in the retro section, which would then be released to leave the capsule to re-enter. Although the re-entry module was similar in size and shape to the Mercury capsule, it had an offset centre of mass to provide a degree of aerodynamic lift and thrusters mounted on its nose to adjust the lift vector so as to steer towards its splashdown point. The only real constraint was that the capsule use the same shape as Mercury, in order to exploit the proven aerodynamics. With a computer and a radar of its own, the Gemini spacecraft would be a much more sophisticated vehicle than its predecessor.

Years of frustration

Following Yuri Gagarin's historic orbit of the Earth in Vostok 1 on 12 April 1961, NASA used Redstone missiles to launch Mercury capsules on suborbital arcs from Cape Canaveral down the Eastern Test Range to splash down in the Atlantic – Al Shepard on 5 May and Gus Grissom on 21 July. The next step for the Soviets was to launch Gherman Titov on 6 August for a day-long mission. This duration reflected the constraint of having the spacecraft land in a zone fairly near the launch site. The Earth rotates on its axis at a rate of 15 degrees per hour and during each of the spacecraft's 90-minute revolutions the planet rotated 22 degrees. After a single revolution, Gagarin had been able to return conveniently. However, on a longer flight, the fact that the ground track slipped westward with each revolution meant that the recovery site would not offer itself again for 24 hours.

On 20 February 1962 John Glenn rode an Atlas and returned after three orbits, which was the most that could be achieved without leaving the narrow zone covered by the tracking and communications sites (this was why three orbits was the official Project target). After Scott Carpenter repeated this on 24 May, it was decided to go for a six-orbit mission, but before this could be attempted the Soviets upped the *ante* considerably.

In early August, Moscow was rife with rumours that a 'space spectacular' was imminent. When Andrian Nikolayev went into orbit in Vostok 3 on the 11th, there was speculation that he would remain up for a week. As his trajectory passed over the launch site 24 hours later, Nikolayev oriented his spacecraft so that he could watch the launch of his colleague, Pavel Popovich. The remarkable consistency in the performance of the Semyorka rocket inserted Vostok 4 into an orbit virtually identical

to that of its predecessor.[2] The precise timing of the launch had been determined after radars at the control centre at Yevpatoria in the Crimea and Ashkabad, near the Iranian border had tracked Vostok 3's approach. This 'group flight' took the world by surprise. When *Tass* announced that the range between the two spacecraft was decreasing, it seemed that the Soviets were about to achieve the Gemini Program's primary objective – even before it could get off the drawing board. Given NASA's decision to develop

On 20 February 1962 John Glenn squeezes into the Mercury spacecraft 'Friendship 7' for America's first orbital mission.

rendezvous as a generic skill, it seemed self-evident that the Soviets would also seek to do so. But appearances were deceptive. The two orbits were similar, but not identical. A vehicle in a low orbit will travel faster than one in a higher orbit. After a brief interval during which the range between the two spacecraft reduced, it had

[2] The Vostoks were launched by the R-7 rocket (the world's first intercontinental ballistic missile) augmented by a small second stage. The name 'Semyorka' means 'old number seven'. Its configuration was so secret that the strap-on configuration was not revealed until one was displayed at the 1967 Paris Air Show. A succession of ever more powerful upper stages have enabled this rocket to be upgraded to launch all manned Soviet spacecraft, and it is still in use today dispatching crews to the International Space Station. In fact, it is the *only* rocket used by the Soviets for manned spaceflight.

Riding an Atlas missile, John Glenn sets off to become the first American to achieve orbit on 20 February 1962.

thereafter increased, and the closest point of approach was about 3 miles. Although official communiques never specifically stated that the spacecraft had *manoeuvred*, Western pundits inferred that they had done so, and the Soviets did nothing to correct this error. NASA did not know it, but apart from firing a retrorocket to return to the Earth, the Vostok spacecraft was incapable of altering its orbit. Nevertheless, its ability to sustain long missions was highlighted by the fact that Nikolayev flew for 94 hours. This dual mission had been developed by Sergei Korolev, the 'Chief Designer' of the Soviet space program, in September 1961, a month after Titov's flight. The failure of a rocket with a military satellite in December, and then the need to make up for this loss, resulted in it being rescheduled from January to March 1962, and then a series of minor issues had delayed it to August. If it had been mounted before John Glenn's mission, it would have been even more spectacular.

On 3 October, Walter Schirra flew NASA's six-orbit mission, and then on 15 May 1963 Gordon Cooper was launched to 'stretch' the Mercury spacecraft to its limit by remaining in orbit for 34 hours, during which, as a succession of the spacecraft's systems failed, he wryly observed that "things are beginning to stack up a little". The next month the Soviets reprised their 'group flight', and this time Valeri Bykovsky – who set a new record of 119 hours and 6 minutes – was fleetingly visited by the first woman in space, Valentina Tereshkova. By now it was evident that NASA stood no chance of catching up until the Gemini spacecraft became available.

Unfortunately, NASA's timetable for the decade slipped due to problems encountered by Mercury, which was drawn out from the envisaged 16 month interval to fully 36 months. In early 1963, NASA hoped to fly the first Gemini test by the end of the year and fly a total of 11 manned missions in the ensuing two years so that Apollo could undertake Earth-orbital tests in 1966, lunar orbital missions in 1967, and land on the Moon in 1968. This unrealistic schedule apart, progress *was* being made.

The line of launch facilities of 'Missile Row' on Cape Canaveral in 1964 includes Atlas and Titan II pads.

The 'boilerplate' spacecraft for the inaugural Gemini mission is hoisted for mating with its launch vehicle on 5 March 1964.

Testing the Gemini launch vehicle

The stages for Gemini 1 were delivered to the Cape in late October 1963, but the task of erecting them on Pad 19 of Missile Row at the Air Force's Cape Canaveral Station took longer than hoped and checking it out dragged on through to the end of January 1964; but it was the first time that a vehicle of this type had been prepared. The 138-foot-tall hinged 'erector' was lowered horizontal and the 71-foot-long first stage with its protruding engine system was placed into it. Then the erector was raised to vertical, and the stage bolted onto the pad structure. Once the erector had been lowered, the 27-foot second stage was inserted, raised and mated. A crane then added the spacecraft to bring the total length of the vehicle to 109 feet. After a flawless countdown, it was launched on Wednesday, 8 April 1964. Upon hearing that it had got off one second late Walter Williams, the Gemini Operations Director, was dismissive: "There must be something wrong with the Range clock!"

Whereas an Atlas departed with a brilliant plume of yellow flame and left behind a billow of white smoke, after an initial blast of thick orange smoke the Titan's hypergolic propellants produced an almost transparent flame, and it rose into the sky as if by magic. Two and a half minutes later, at an altitude of 35 nautical miles, the first stage shut down. The second stage's engine lit immediately (a procedure referred to as a 'fire in the hole' ignition) and pushed the spent stage clear. Upon orbital insertion 540 miles down the Eastern Test Range, the 87 mile perigee was perfect, but a 24 foot per second overspeed meant a slightly higher than planned apogee. The spacecraft was only a 'boilerplate' model that did not separate from the rocket stage, so the entire stack burned up when it re-entered the atmosphere four days later. This test had verified the compatibility of the launch vehicle and the spacecraft during the ascent phase.

The first Gemini-Titan lifts off from Pad 19 on 8 April 1964.

A month earlier, Williams had announced his resignation, which was to take effect upon the dispatch of this test. He was superseded as Gemini Operations Director by Chris Kraft, who had been serving as Assistant Director for Flight Operations at the Manned Spacecraft Center in Houston since November 1963.

Initially, only one unmanned test had been envisaged, but when he took over as Program Manager in March 1963 Charles Mathews had ordered a second test using a fully-functional spacecraft. At that time, it was hoped to launch Gemini 2 six months after the first test, but several factors conspired to delay it.

The Soviet bluff

In August 1964, Khrushchev made achieving a lunar landing ahead of the Americans a national priority. Two months later, NASA was shocked by the launch of Voskhod 1, which had a three-man crew. Could this be the Soviet version of Apollo? So soon? Could they really be *that* far ahead? The use of a new name was designed to suggest that a new spacecraft was being tested, but in fact it was a Vostok, and the only way to squeeze in the command pilot Vladimir Komarov, spacecraft engineer Konstantin Feoktistov, and medical specialist Boris Yegorov was to replace the ejection seats with lightweight couches and have them fly without pressure suits, which meant that in the event of a launch mishap they would have no means of escape. During this 24-hour 'stunt' Khrushchev was deposed by Leonid Brezhnev.

Vladimir Komarov (left), Boris Yegorov and Konstantin Feoktistov walk in their comfortable flight suits to their Voskhod 1 capsule on 12 October 1964.

The Gemini 2 spacecraft is hoisted up for mating with its launch vehicle.

Man-rating the Gemini spacecraft

Gemini Launch Vehicle 2 was delivered to the Cape on 10 July 1964, erected on Pad 19 on 14 July, and its check-out begun. But when lightning struck the pad on 17 August it was lowered for a thorough inspection. A fortnight later, the Cape was threatened by Hurricane Cleo and the upper stage was retrieved and placed in storage; the first stage was left on the pad. No sooner had the stage been remated than Hurricane Dora loomed.[3] This time, it was decided to remove the entire vehicle. Dora turned away from the Cape, but Ethel was close behind. Consequently, the vehicle was not returned to the pad until mid-September. Testing restarted on 21 September and ran through to 6 October. Meanwhile Spacecraft 2 had been delivered and also slipped behind schedule, and was not able to be mated until 5 November, which marked the start of a month of integration checks to verify that the 'automated pilot' functioned properly.

At 11:41 local time on Wednesday, 9 December, as Gemini 2's countdown reached its conclusion, the Titan's engines ignited – and promptly shut down again. The Malfunction Detection System that NASA had added to the launch vehicle specifically for Gemini had noted a fall in pressure in the hydraulic system that would gimbal the engine to steer the vehicle in flight, causing it to swing 'hard over' to its limit. Although the backup system stepped in to regain control of the engine, the MDS intervened and shut down the engine because it was programmed not to launch on a backup system, as this would mean no in-flight redundancy. This 'pad abort' was feasible

[3] As a weatherman said, warning of Hurricane Dora, "If you thought Cleo was a honey, look out for this one."

because the vehicle had not yet lifted off. In fact, the signal to fire the pyrotechnic 'hold back' bolts that held the base of the Titan to the pad was not due to be issued until three seconds beyond T=0, and it was conditional upon both chambers of the engine system having successfully run up to full thrust.[4] Gus Grissom and John Young, and their backups Wally Schirra and Tom Stafford, had flown down to observe the launch. They were in the blockhouse. Bastian Hello, one of the Martin Company's managers, was observing through a periscope. He turned to the astronauts and said in amazement, "It's still sitting there!" While the 'fizzle' was frustrating, the fact that the MDS had save the vehicle was encouraging. "The fact is," reflected Grissom later, "the abort taught us something: our MDS worked."

Once the engine had been serviced, the launch was set for 19 January 1965, and it lifted off without incident. In contrast to Mercury, the Gemini spacecraft was "a pilot's vehicle" with the minimum of automation, so for this test flight it carried an electromechanical sequencer to activate the controls at the appropriate times. Within minutes of separating from the upper stage, the sequencer fired the OAMS attitude control thrusters to yaw to what was referred to as

Because Gemini was designed to be flown by human pilots, an electro-mechanical sequencer had to be installed for the flight that would 'man rate' the spacecraft.

'big end forward' attitude flying backwards, jettisoned the adapter module, ripple-fired the four solid-fuel rockets in the retro module to de-orbit itself, and shed the retro module to expose the re-entry capsule's heat shield. The spacecraft splashed down in the Atlantic, just beyond the end of the Eastern Test Range, where it was recovered by a helicopter from the *USS Lake Champlain*.[5] Although it lasted a mere 18 minutes, this flight had 'man rated' the spacecraft.

---

[4] Without a malfunction detection system, the Atlas would not have been able to shut itself down and it would have lifted off running on its backup system; if it had one.

[5] Although commonly described as suborbital, this spacecraft's trajectory was not the same as the ballistic arc flown by the early Mercury missions; it simply de-orbited itself before completing one orbit. It would better be described as a 'fractional orbit' trajectory.

As the first stage of Gemini Launch Vehicle 2 ignites on 9 December 1964, the Malfunction Detection System intervenes to shut the engine down.

By this point, the Program was two and a half years behind schedule and, at a projected cost of 1.3 billion dollars, was likely to consume twice its original budget. However, this did not so much indicate that it was running out of control, rather that the initial plans had been unrealistic.

The Soviets score again

Following a preliminary study in 1963 of a cosmonaut exiting his spacecraft to undertake a 'spacewalk', Korolev was instructed in March 1964 to arrange for one member of a two-man crew to temporarily exit the spacecraft, in order to upstage NASA, which had announced that a Gemini astronaut would do so. The development of an extravehicular spacesuit and the necessary airlock took rather longer than hoped, and the mission slipped into the new year. Soon after Voskhod 2 attained orbit on 18 March 1965, Alexei Leonov inflated the airlock and went through it to become the first man to 'walk' in space. Pavel Belyayev remained inside. Leonov was testing a suit intended for a lunar mission, in which a crewman would transfer from one spacecraft to another. Because it had an independent life-support system, he was tethered only as a safety measure. The official communiques did not mention that his suit 'ballooned up' in vacuum, and he had to partially deflate it in order to get back into the airlock. Nor was it reported that after the planned 24 hours the automatic retro system failed, and after an additional revolution Belyayev had to manually trigger the backup package, did so late, and caused them to land on a mountainside and endure a night in the snow before being rescued.

Meanwhile, at Cape Kennedy,[6] the final preparations were being made to dispatch the first Gemini crew.

---

[6] After the assassination of John F Kennedy on 22 November 1963, Cape Canaveral in Florida became Cape Kennedy, and the NASA facility there was named the Kennedy Space Center.

The early chronology

Spacecraft	Launched	Crew	Duration	Revs
Vostok 1	12 April 1961	Yuri Gagarin	1h 48m	1
Freedom 7	5 May 1961	Al Shepard	15m 22s	0
Liberty Bell 7	21 July 1961	Gus Grissom	15m 37s	0
Vostok 2	6 August 1961	Gherman Titov	25h 18m	18
Friendship 7	20 February 1962	John Glenn	4h 55m	3
Aurora 7	24 May 1962	Scott Carpenter	4h 56m	3
Vostok 3	11 August 1962	Andrian Nikolayev	94h 0m	64
Vostok 4	12 August 1962	Pavel Popovich	70h 50m	48
Sigma 7	3 October 1962	Walter Schirra	9h 13m	6
Faith 7	15 May 1963	Gordon Cooper	34h 20m	22
Vostok 5	14 June 1963	Valeri Bykovsky	119h 6m	81
Vostok 6	16 June 1963	Valentina Tereshkova	70h 50m	48
Voskhod 1	12 October 1964	Vladimir Komarov Konstantin Feoktistov Boris Yegorov	24h 17m	16
Voskhod 2	18 March 1965	Pavel Belyayev Alexei Leonov	26h 2m	17

Three frames from a newsreel released by the Soviets
of Alexei Leonov's historic spacewalk from Voskhod
2 on 18 March 1965.

Chapter 2

Shakedown Flight
Gemini 3

Objectives

Preparations to launch Gemini 3 began a week after the second test flight, with the erection of the Titan II launch vehicle on Pad 19 on 25 January 1965, and the addition of Spacecraft 3 on 17 February. For once, events progressed so smoothly that NASA was able to advance the mission from early in April to 23 March. The flight plan was finalised on 14 March. The next day, Gus Grissom and John Young flew to the Cape to participate in the Flight Readiness Review on 18 March,[1] and were themselves cleared by a medical review on 21 March.

The objectives of this 'shakedown' flight, as it was being described, had been a matter of debate for two years. When Charles Mathews took over from James Chamberlin as Program Manager in March 1963 the plan had been for the first manned Gemini mission to match Gordon Cooper's final Mercury flight, which was scheduled for May, but in April it was cut to three revolutions, which was the most that it could fly before the Earth's rotation would cause its ground track to depart from the coverage of the World-Wide Tracking Network – Mathews wanted the spacecraft to be able to be monitored throughout its flight. Nevertheless, in the summer of 1964 the Astronaut Office argued for an 'open ended' mission that would remain in orbit either for as long as Mercury's 34-hour record or until such time as a fault obliged a recall, but this was ruled out. The *gung ho* attitude of the astronauts contrasted markedly with the nervousness of the managers – and the higher the level of seniority, the greater the degree of caution. At this stage in the Program, the most important objective was to confirm that the spacecraft could modify its orbit by firing its OAMS thrusters, and a series of tests were planned to demonstrate this. In January 1965, however, NASA Headquarters ordered that a 'burn' be added to the flight plan to guarantee that the spacecraft would re-enter the atmosphere in the event that its solid-propellant retro rockets failed.[2] The directive that was issued on 15 February 1965 described Gemini 3's mission as "to demonstrate and evaluate the capabilities of the spacecraft and launch vehicle system, and the procedures necessary for the support of future long-duration and rendezvous." Although it was not evident how such a *brief* mission would significantly contribute to long-duration studies, *manoeuvring* in space would be a major step towards orbital rendezvous.

---

[1] As it happened, Gemini 3's Flight Readiness Review was conducted on the same day as Alexei Leonov made the Soviet Union's pioneering spacewalk.

[2] This 'fail safe' burn was apparently prompted by the scenario depicted by Martin Caidin in his 1964 novel *Marooned*, in which a spacecraft became stranded in orbit when its retros failed to fire.

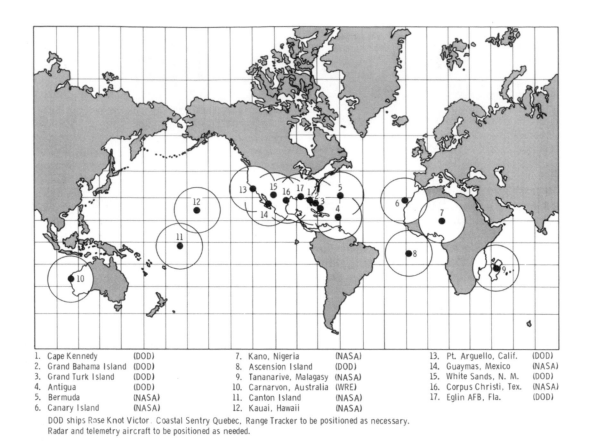

1.	Cape Kennedy	(DOD)	7.	Kano, Nigeria	(NASA)	13.	Pt. Arguello, Calif.	(DOD)
2.	Grand Bahama Island	(DOD)	8.	Ascension Island	(DOD)	14.	Guaymas, Mexico	(NASA)
3.	Grand Turk Island	(DOD)	9.	Tananarive, Malagasy	(NASA)	15.	White Sands, N. M.	(DOD)
4.	Antigua	(DOD)	10.	Carnarvon, Australia	(WRE)	16.	Corpus Christi, Tex.	(NASA)
5.	Bermuda	(NASA)	11.	Canton Island	(NASA)	17.	Eglin AFB, Fla.	(DOD)
6.	Canary Island	(NASA)	12.	Kauai, Hawaii	(NASA)			

DOD ships Rose Knot Victor, Coastal Sentry Quebec, Range Tracker to be positioned as necessary.
Radar and telemetry aircraft to be positioned as needed.

Although ships were used to fill gaps in the World-Wide
Tracking Network, spacecraft were not in constant communication.

'Molly Brown'

Having named his Mercury spacecraft '*Liberty Bell*' and painted a crack across its outer shingle casing, only to have it sink on him upon splashdown, Grissom had decided to call his new vehicle after the heroine of the popular Broadway show '*The Unsinkable Molly Brown*'. When Washington expressed misgivings and asked Grissom to come up with something else, he offered '*Titanic*'. Although James Webb belatedly accepted *Molly Brown*, he ordered an end to the practice of naming spacecraft.[3]

On Monday, 22 March, Grissom and Young had supper in the Astronaut Quarters in the Manned Spacecraft Operations Building on Merritt Island with Wally Schirra and Tom Stafford (their backups), Al Shepard, Deke Slayton, Jim McDivitt and Ed White (the next crew in line) and an old friend Leo DeOrsey. As they ate, the Martin Company's team began to pump propellant into the Titan's two stages. On being awakened by Slayton at 04:40 local time they had a traditional pre-flight steak and eggs breakfast with all the trimmings prepared by Sam Piper, and were joined by

[3] This came as a disappointment to the Gemini 4 crew, Jim McDivitt and Ed White, who wanted to name their spacecraft '*American Eagle*'.

Slayton, Shepard, Chris Kraft, who was serving as both the Mission Director and Flight Director; Dr Charles Berry, the chief Flight Surgeon; Charles Mathews, the Program Manager; Walt Williams, until recently the Program's Operations Director; Robert Gilruth of the under-construction Manned Spacecraft Center in Houston; Merritt Preston, the Deputy Director of the Kennedy Space Center; James 'Mr Mac' McDonnell and Walter Burke of McDonnell Aircraft, the spacecraft manufacturer; Bastian Hello of the Martin Company which supplied the Titan II; and Air Force Colonel Richard Dineen, who managed the Titan launch facilities.[4] In a moment of levity, Young, who had grown up in nearby Orlando, was given a 60-foot-long telegram from 2,400 residents of that city wishing "their boy" a good flight.

On arriving at the suit-up trailer on Pad 16, Grissom and Young were met by Wally Schirra in a battered old Mercury suit.

Schirra and Stafford spent the night in a van at Pad 16 on Missile Row, which had been built to launch the Army's Pershing missile and was now disused. It was a few hundred feet south of Pad 19, the Titan II facility that was configured for the Gemini Launch Vehicle. They awoke early in order to configure the spacecraft. The 'White Room' was cleared for 45 minutes as a precaution while nitrogen was pumped into the Titan's propellant tanks to raise them to their operating pressure.

At 06:00 a two-car motorcade took Grissom and Young out to Pad 16, where a trailer had been set up to enable them to don their pressure suits. They were met by Schirra, wearing a Mercury suit that had seen better days, who announced that he was ready "just in case you two chicken out". After being fitted with sensors for biomedical telemetry, Grissom and Young were helped into their suits by technicians Al Rochford and Joe Schmidt. They sealed their visors to breathe pure oxygen in order to purge nitrogen from their blood stream so as to be able to reduce the suit pressure once in space without suffering 'the bends'. Until they could plug into the spacecraft they were to use portable ventilators, which they would carry in the manner of suit cases. At 07:06, the four astronauts boarded the transfer van to drive the new road that had been laid directly from Pad 16 to Pad 19. This saved a roundabout route of almost a mile. It had been ordered by Air Force warrant officer 'Gunner' Barton, and so had been dubbed the 'Barton Freeway'. After disembarking, they walked up the ramp to the red-metal erector, rode the wire-frame elevator

---

[4] Although NASA's Cape facility had been renamed the John F Kennedy Space Center, the Titan II pad was on the Air Force's Canaveral Station.

Gus Grissom suits up to command the first manned Gemini mission.

100 feet up to the White Room level, where they were received by Guenter Wendt, the 'Pad Leader' of the McDonnell ingress team. Compared to the 'tight' hatch of the Mercury capsule, the Gemini spacecraft's large swing-open hatches made ingress straightforward – 'sit' in the recumbent seat with feet straight up, strap on the harness, transfer the umbilical from the ventilator to the spacecraft's environmental control system, and then hook up the radio. In keeping with tradition, Wendt shook each astronaut's hand warmly immediately prior to closing his hatch. The cabin was pumped up with pure oxygen, and the hermetic seals of the hatches verified. At 07:34 Stafford left for the pad's blockhouse to serve as 'Stoney' CapCom, and Schirra went to what had been 'Mercury Control' and which, despite having been updated to serve as 'Gemini Control', was soon to be superseded by the 'Mission Control' at the Manned Spacecraft Center in Houston.[5]

Although the count had been 20 minutes ahead of nominal when the crew had arrived on Pad 19 and the various events were progressing smoothly, a low overcast threatened a delay. When a leak was noted in the oxidiser line to the first stage of the launch vehicle, a 'hold' in the countdown was called at T–35 minutes, and a technician with a wrench set off from the blockhouse to make a visual inspection; he stemmed the leak by tightening a loose nut. The count resumed at 07:45, the pre-valves on the first stage were opened to permit the oxidiser to 'condition' the lower part of the engine system, the spacecraft was transferred to internal power, and the 138-foot-tall erector was slowly rotated down from vertical to horizontal, to leave Gemini 3 standing alongside its umbilical tower.

At T–20 minutes the Spacecraft Test Conductor, George Page, ordered the test of the 25-pound-thrust attitude control thrusters around the rim of the spacecraft's

---

[5] In fact, one of the Mission Operations Control Rooms in Houston's Manned Spacecraft Center was playing a passive monitoring role for this mission, in order to test its systems before assuming prime status for the next mission.

The 138-foot-tall erector begins to swing down from Gemini 3.

adapter, with each being briefly 'blipped'. In addition to confirming that the thrusters worked, this primed the propellants in the feed pipes. A poll of the flight controllers at T–10 minutes yielded a 'Go' for launch and this word was passed to the spacecraft by Gordon Cooper, the CapCom in Gemini Control. Even the overcast had cleared. At T–2 minutes, the pad was deluged with water in order to cool the structure from the hot engine exhaust. The first stage's engines, which projected far below the cylindrical tankage, were moved to verify the full range of motion of the gimbal that was to create the thrust-vectoring to steer the vehicle in flight. The pre-valves were opened to permit the fuel to enter the engine system – each of the propellants was now only one valve away from the injectors. The ignition command was issued at 09:24. After 3.4 seconds, by which time the twin-chambered engine was confirmed to have run up to at least 77 per cent of its nominal 430,000 pounds of thrust, the pyrotechnic hold-back bolts were fired and the Titan left the pad. As it did so, a plug was pulled out of the tail of the first stage and this action started the spacecraft's mission event timer.

"The clock has started," Grissom announced. It was a very smooth departure, so smooth in fact that neither man had any sensation of motion. It was a contrast to the Redstone that Grissom had ridden for his Mercury flight. Despite the thunderous roar that washed over the Cape, it was quiet inside the cabin. "There's the roll program," Grissom added as the Titan rolled to align its inertial guidance system. They felt the slight lateral shifts as the engine 100 feet below was gimballed to maintain the vehicle upright. "Pitch program." Once aligned on the appropriate azimuth, the rocket began to pitch over to fly an arc down the Eastern Test Range.

"You're on your way, *Molly Brown*!" observed a delighted Cooper.

"Yeah, man!" Grissom agreed. It was a milestone for him because he was finally heading into orbit. If his suborbital Mercury mission was counted, then he was the first human being to set off for space twice, so even though the mission had barely begun NASA was already able to claim a 'first' over the Soviet Union.

Grissom kept a grip of the 'D'-ring, ready to eject if the Malfunction Detection System indicated a non-recoverable fault in the vehicle. Young ignored his 'D'-ring, content to leave this call to his Command Pilot.

"Plus 50 seconds," Cooper advised.

"Roger, Mode II-delayed," Grissom acknowledged. They had exceeded the ceiling for the safe use of the ejection seats – Mode I. If the MDS signalled a fault and shut down the Titan, Grissom would wait for five seconds, then separate the re-entry and retro modules from the stack, rotate 'blunt end forward' (BEF), and fire the four solid retro-rockets to fly a ballistic arc for recovery on the Eastern Test Range, in a replay of his Mercury mission.

As the vehicle went supersonic, the increasing aerodynamic pressure rattled the structure. At 1 minute 20 seconds, as it passed through 40,000 feet, this rattle reached its peak and then rapidly tailed off.[6] "It just got quiet," Grissom observed.

"One plus 40," Cooper advised.

"Roger, Mode II." If an abort was required now, Grissom would act immediately rather than wait five seconds.

"First DCS update," Young confirmed, when the lamp on the Digital Command System illuminated to show that the Cape had sent a navigational update, based on tracking of their ascent by the Cape's radars.

"You're a little bit high on the flight path," Cooper pointed out, "but it's no problem."

"*Molly Brown* is 'Go' for staging," Grissom advised. By this time, the acceleration had risen to 3.3 *g*.

"Second update received," Young noted.

After two and a half minutes the first stage shut down. The 'fire in the hole' ignition of the second stage's single-chambered engine blasted efflux out through a ring of vents in the interstage, creating a characteristic ring of fire. To the crew, for whom the sky had already turned black, the brilliant orange-yellow flash was startling. The spent stage was released by explosive bolts and blasted downward by the efflux from the second stage as it accelerated away.

"There was staging," Grissom reported, "and we're thrusting." The second stage started to manoeuvre to correct the steep trajectory that it had inherited from the booster. "We're starting to steer."

If the MDS signalled a Mode III abort now, as soon as the Titan's engine had been shut down Grissom would fire the pyrotechnics around the rim of the adapter module to release the spacecraft and separate using the two 100-pound-thrust aft-firing OAMS thrusters, as a preliminary to executing the normal de-orbit procedure leading to recovery far downrange in the Atlantic.

"*Molly Brown*, you're 'Go' from here," Cooper advised as they approached the end of the second stage's thrust. "You're steering right down the line." The indicator on the display in the control room was following the arc for the required ascent. "Point-8," Cooper called. Having attained 80 per cent of the speed required for the nominal orbit, if the second stage shut down prematurely Grissom would be able to separate the spacecraft and limp into low orbit utilising the OAMS thrusters. As the second stage consumed its propellants it became lighter and rapidly accelerated, with the load increasing to a peak of 7.2 *g*. As a consequence of the fact that the Titan II rolled early in the ascent to align its inertial guidance system, the spacecraft was

---

6 The maximum aerodynamic pressure (referred to as 'Max Q') was 750 pounds of force per square foot.

carried 'right side low', and whereas Grissom saw only black sky out of his tiny 'half-moon' window, Young had an awe-inspiring view down the Eastern Test Range.[7] However, when the upper stage made a final refinement of the trajectory, the spacecraft's nose momentarily dipped below horizontal and Grissom got his first visual confirmation of their attitude.

"SECO!" Grissom called as the second stage's engine cut-off after 3 minutes of smooth thrusting. In weightlessness, bits of detritus promptly emerged from the nooks and crannies of the cabin and floated around.

As some of the missile versions of the Titan II had exploded soon after dispatching their payloads, the flight plan called for the spacecraft to separate from the upper stage within 30 seconds of shutdown. Young threw the switch to fire the pyrotechnics (which he later said sounded "like the bark of a battery of howitzers") and Grissom fired the aft-facing OAMS thrusters for 15 seconds to move well clear.

Cooper reported that radar tracking indicated that they were in an orbit ranging between 87 and 125 nautical miles. The spacecraft's Incremental Velocity Indicators (IVI) had shown that the Titan had produced a 17 foot per second overspeed. The separation manoeuvre had increased this to 29 feet per second. Roughly speaking, each 2 foot per second produced a 1 mile increase in apogee, so the elliptical orbit was slightly more eccentric than intended, but this would not pose a problem.

The first task was to run through the post-insertion checklist to verify the systems, and to configure the spacecraft for orbital flight.

As they neared the far end of the Eastern Test Range, Cooper posed a personal question to Grissom. "Does it look better than on a ballistic flight?"

"Say again?" Grissom requested. Cooper did so, but the communications were fading. "I can't read you, Gordo."

"HOW DO IT LOOK?" Cooper articulated loudly and clearly.

"It look great!" Grissom replied enthusiastically. Their altitude was similar to that at the top of his suborbital arc produced by the Redstone missile, but the view was better.

"Rog," Cooper signed off.

Flying in space

As they headed across the Atlantic, Grissom and Young worked through their checklist. When the Canary Island communications station made contact, it read up the data for a re-entry that would result in a splashdown at the prime recovery site in the Atlantic at the start of the second orbit. This was a contingency measure to enable the spacecraft to return to Earth safely if the communications system failed completely. Young wrote the information into a blank entry in the flight plan and then read it back for confirmation. Grissom reported their first problem: "I seem to have a leak in one of the thrusters, because I get a continuous yaw to the left." At a quarter of a degree per second, this yaw was annoying, but it could be readily overcome when the spacecraft had to be held in a specific attitude. His report set the engineers at

---

[7] Each of the 'half-moon' windows was only 8 inches wide and 6 inches deep.

McDonnell thinking as to the nature of the fault.

Several minutes after leaving Canary behind, just as he finished the checklist, Young saw the indicator on the oxygen pressure gauge suddenly fall. "We've lost ..." he began, but then he saw that several other instruments were giving anomalous readings, and realised that the fault involved the power supply to the instruments. When he switched to the backup power supply the affected instruments resumed their nominal values. Within 45 seconds of spotting something amiss, the engineering test pilot had diagnosed and rectified the fault.

"What have we lost?" Grissom asked.

"A primary converter."

"Really?" On observing the falling oxygen pressure, Grissom had closed his visor to seal his suit, so he rather sheepishly opened it again and grinned – if the environmental system's oxygen supply had failed, sealing his visor would not have done any good.

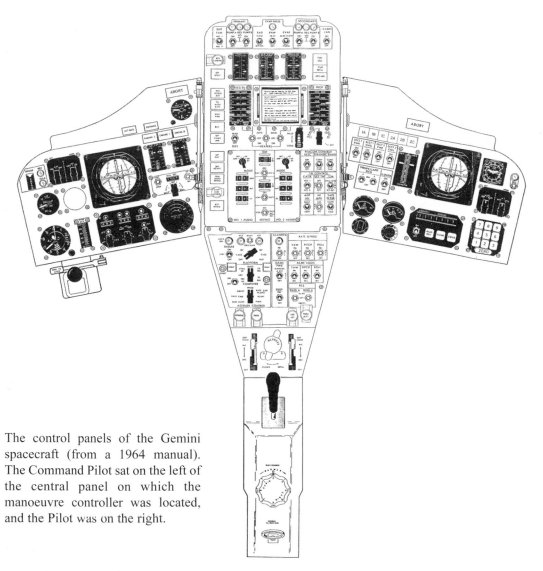

The control panels of the Gemini spacecraft (from a 1964 manual). The Command Pilot sat on the left of the central panel on which the manoeuvre controller was located, and the Pilot was on the right.

Moments later, Cooper announced contact via the voice-relay station at Kano in Nigeria, and asked about the yaw problem. Grissom replied that they were still drifting but it was not serious, and then told Cooper of their new problem: "We lost our primary dc-dc converter." Cooper acknowledged this and passed on some tests that the engineers wanted them to make to help to characterise the yaw issue. Flying on across Africa, Gemini 3 crossed the evening terminator, which was the line dividing the sunlit and dark hemispheres. As they flew over a storm system, Grissom noted some flashes of lightning. Communications via the Tananarive relay on the island of Madagascar were poor. Flying on in darkness, they performed the first of a series of measurements of blood pressure and temperature to monitor how their bodies were adapting to weightlessness. On establishing contact with the *Coastal Sentry Quebec*, a former 'Liberty Ship' serving as a communications ship on station in the Indian Ocean, they were asked to conduct further tests to diagnose the nature of the mysterious left yaw. Pete Conrad was the CapCom at Carnarvon on the west coast of Australia. As Gemini 3 climbed to apogee, it rose over his horizon. He cleared them to go ahead with the second revolution, and relayed the data for the orbital adjustment that was to be performed when they returned to perigee. Cooper made contact via the relay site on Canton Island, between Australia and Hawaii, and advised them that the Texas station would continuously monitor the propellant pressures in an effort to estimate the rate of depletion from the suspected leaky thruster.

"Look at the sunrise!" exclaimed Grissom as they approached the dawn terminator.

"Isn't that beautiful!" Young agreed.

"Aren't you going to take any pictures?"

The flight plan did not call for a picture at this point, but the sight was so spectacular that Young retrieved his camera and took a few 'tourist shots'.

The *Rose Knot Victor*, which was on station in the northeast Pacific, provided an update on the forthcoming manoeuvre and gave the go-ahead to undertake it.

"We're in good shape," Grissom assured as they left the ship behind.

A few minutes later, North America hove into view. "Outstanding!" Young exclaimed in astonishment. "I can see the whole of Mexico."

The CapCom at the Guaymas site on the mainland coast of the Gulf of California handed the spacecraft over to his counterpart at the Corpus Christi station in Texas, who monitored the propellant pressures.

As Gemini 3 approached perigee, Grissom cancelled the yaw drift to hold the spacecraft 'small end forward' (SEF) for the burn. "Mark!" he reported, when he ignited the forward-firing thrusters for the 74-second burn to slow the spacecraft and thereby lower its apogee.

"They appear to be firing good," Young observed; there was no noise, but the squirting efflux was visible out of the corner of his window.

"Thrusting complete," Grissom reported. The desired velocity change had been manually entered into the IVIs, and Grissom had burned these meters down to zero. The thrusters had performed as expected and, despite its tendency to yaw, the spacecraft had held its attitude well. The manoeuvre had altered the initially eccentric

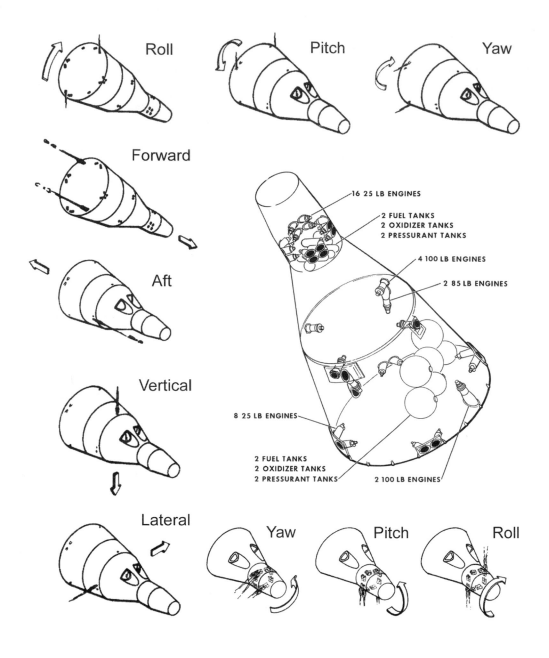

Roll

Pitch

Yaw

Forward

16 25 LB ENGINES

2 FUEL TANKS
2 OXIDIZER TANKS
2 PRESSURANT TANKS

4 100 LB ENGINES

2 85 LB ENGINES

Aft

8 25 LB ENGINES

Vertical

2 FUEL TANKS
2 OXIDIZER TANKS
2 PRESSURANT TANKS

2 100 LB ENGINES

Lateral   Yaw   Pitch   Roll

A pair of 100-pound-thrust rockets in the rear provided forward translation and a quartet of such thrusters positioned around the middle directed their thrust through the vehicle's centre of mass for lateral translations. A pair of forward-firing 85-pound rockets served to reverse. Eight 25-pound rockets around the rim of the adapter were fired in differently selected pairs for roll, pitch and yaw. On the nose of the re-entry module were two 'rings', each with eight 25-pound rockets, for roll, pitch and yaw after the adapter had been jettisoned for re-entry. (Not shown are the four solid rockets in the retro section for the de-orbit manoeuvre.)

orbit to one that was nearly circular, ranging between 86 and 93 nautical miles. It was a historic moment for manned spaceflight.

Cooper advised that when they reached their second sunset they would be close to their spent Titan stage. They should be able to see it dead ahead, reflecting the Sun. "Proceed to see if you can rendezvous" with it, he jokingly suggested. In the VIP enclosure, Gilruth and his deputy George Low heard this and decided that it would make an interesting experiment on the next mission.

Shortly after completing their second pass over Canary, Young pulled a surprise packet from his suit pocket and offered it to his commander.

"What is it?" Grissom asked.

"A corned beef sandwich."

"Where did it come from?"

"I brought it with me," Young said. "See how it tastes," he prompted. The aroma was enticing.

Grissom took a bite. "It's breaking up. I'm going to stick it in my pocket." He did not want crumbs floating in the cabin.

"It was a thought, anyway," Young reflected. "Want some chicken bites?" he offered as he selected one of the packets on the list of *bona fide* foodstuffs that they were to evaluate.

"No, you can handle that," Grissom prompted. As he listened to Young chomp through the meal, he was moved to remark "You're a noisy eater!" Young pulled out a bag of apple sauce and squeezed some of the goo out through the nozzle. It met with his approval. When Grissom asked for some he, too, was enthusiastic: "Now if we had some pork chops to go with it, we'd be alright!"

As they approached their second sunrise, they looked for the Titan stage, but could not see it. In darkness, in range of the *Coastal Sentry Quebec*, Grissom yawed the spacecraft to one side and fired first the forward-facing thrusters for 10 feet per second and then the aft-facing thrusters in an effort to cancel out the first burn, which he was almost able to do, with the 1 foot per second residual slightly increasing the inclination of their orbit with respect to the equator – these manoeuvres had been made 'out of plane' in order not to disturb their circular orbit.

At Carnarvon, Conrad speculated on how a leaky thruster might be causing the observed yaw drift. "It sounds to me like you've got a mechanical problem in the valve." The thruster incorporated a valve that was operated by a solenoid.

"It must be very slight," Grissom noted, "because we cannot see the pressure go down." The yaw thrusters were arranged around the rear periphery of the adapter, and were fired in pairs. If one was leaking, then there ought to be a degree of 'coupling' that would induce a roll component, but this was not evident. When Conrad began to offer further tests, Grissom was insistent, "We've gone through *everything*, Pete!"

Upon breaking into daylight, they were temporarily blinded by the harsh Sun. This time their track over the Pacific reached Hawaii, where Neil Armstrong was on duty as CapCom. Grissom reported that the Flight Director Attitude Indicator (which was also referred to as the '8-ball') was "drifting badly" because it was 10 degrees off – either that, or there was a fault in the system which was meant to rotate the spacecraft through 360 degrees in order to match its orbital rate and thereby maintain its orientation with respect to the Earth. Because each pilot had his own '8-ball', Armstrong enquired whether these were in step; they were. "The Orbit Rate isn't any good," Young concluded.

The *Rose Knot Victor* requested a comprehensive status check, in which the astronauts read out the onboard gauges so that their readings could be checked against the telemetry.

"How's the weather around the world?" Cooper asked, once the Cape picked them up.

"Very cloudy!" Grissom noted.

"I see."

"We've seen very little land."

"All clouds and water, huh?" Cooper commiserated.

"Not even much water!" Grissom lamented.

As Gemini 3 set off across the Atlantic for the third and final time, Cooper read up the data for the final manoeuvre.

"The pre-retro checklist is complete, Pete," Grissom reported when they made contact with Carnarvon.

"Gus," Conrad began, "I only have one question for you before you go out of range, how is the *flying* up there?"

"Great!!"

As they crossed the Pacific for the final time, Grissom and Young stowed the loose items away. The 'fail safe' manoeuvre was made a few moments after contacting Hawaii using the aft-facing thrusters while flying BEF and 'heads down'. This lowered the perigee of the orbit to about 45 miles, thereby guaranteeing that they would re-enter the atmosphere if the retros failed.[8]

"The adapter has separated," Grissom informed the *Rose Knot Victor* as the equipment section of the adapter module was jettisoned. The explosive bolts which blasted it free were violent, but the remainder of the spacecraft held its attitude. The ship's CapCom read up the count ending in "Retrofire!" Grissom reported the firing of each of the four solid-rockets. A minute later, now in contact with Guaymas, Young reported that they had also jettisoned the retro section of the adapter. Cooper monitored as the re-entry module flew east, with its heat shield facing the direction of travel, down to the entry interface at an altitude of 400,000 feet.

A spacecraft returning from the Moon would enter the Earth's atmosphere considerably more rapidly than one from orbit, and its heat shield would have to endure correspondingly greater thermal stress. This posed a challenge for the designers of the Apollo mothership. One option was to slow down by striking the atmosphere at a shallow angle in order to 'skip' off and pursue a ballistic arc that would result in re-entry. Considering that the Mercury capsules had come down far from their targets, the very accurate navigation required for such a scheme appeared daunting. Gemini was therefore to develop techniques for 'controlled re-entry'. The

[8] Of course, once the retros fired the spacecraft would never *reach* this notional perigee because the atmosphere would intervene.

centre of mass of the re-entry module had been offset in order to provide the blunt cone with some aerodynamic lift. By monitoring the trajectory, the onboard computer could predict the splashdown point and display this to Grissom, who could adjust the 'lift vector' by using the thrusters in the nose to roll left or right of the 'neutral' position, and thus steer towards the target.

"Hello there!" Cooper welcomed on finally regaining contact as the capsule fell through 80,000 feet.

"The needles show us about 25 miles short," Grissom reported. The 'needles' provided the computer's projection of their trajectory. When this indicated that they were coming in 'short' he had attempted to 'extend' but had been unable to do so, and they came down 45 nautical miles from the *USS Intrepid*, which was on station in the Atlantic near Grand Turk Island in the Bahamas. Subsequent analysis revealed that the re-entry module had less lift than believed, but in the days before high-speed computers, hypersonic flight characteristics were difficult to estimate because wind tunnels could not generate such speeds. Nevertheless, the role of an *engineering test flight* was to determine the vehicle's performance, and this empirical data could readily be taken into account on future missions.

As they descended, waiting for the drogue chute to deploy, Grissom was primed to pull his 'D'-ring to eject if it failed to do so. The big surprise on this brief but impeccable mission was the main chute. After the drogue pulled it out, Grissom threw a switch to transition from a single point of attachment with the capsule's nose straight up to a two-point configuration in which it was angled slightly above the horizontal. However, this transition was so violent that, even though he was strapped in, Grissom's head was thrown forward and his faceplate was shattered when it struck the frame of his window.

After the capsule hit the water, the parachute dragged it under water before Grissom was able to throw the switch to jettison it, and for a moment all he could see through his window was green water – which brought back memories. Once it was free, however, the spacecraft bobbed upright.

"This is *Molly Brown* in the water," Grissom called. "Anybody read?" The reply by the recovery aircraft circling overhead was immediate. Upon being told that the ship was so far away he asked for a helicopter pick-up. In light of his *Liberty Bell* experience he refused to open the hatches until the swimmers had the flotation collar in place, and so he and Young suffered the discomfort of pitching and rolling in the 5-foot seas. As Young reflected later, "Gemini may be a good spacecraft, but she's no boat." Having both vomited, they stripped off their suits. Because the capsule had settled in a slightly lopsided position, they exited by the left hatch, which prompted Young, a naval officer, to wryly observe that it was the first time that he had seen a commander be first to abandon his ship. On being hoisted on board the helicopter, they put bathrobes over their longjohns. Seventy five minutes after splashing down, they reached the *USS Intrepid*, where they were given a red carpet welcome. After a battery of medical tests, they took a telephone call from President Johnson and, while their memories were fresh, relived the mission in the first of a series of debriefings. The next day they were flown back to the Cape for a heroes welcome.

Having changed out of their soiled suits and donned bathrobes, Grissom and Young cross the deck of the *USS Intrepid*.

The swimmers stay with *Unsinkable Molly Brown* until the prime recovery ship can hoist it aboard.

Gus Grissom, President Lyndon Johnson and NASA administrator James Webb.

"The entire flight was completely satisfactory," proclaimed Chris Kraft at the post-flight press conference, "just about perfect." Gemini was the first spacecraft capable of *flying* in space. This test flight had certified the cabin's environmental system, the pilot's instrument displays and hand controller, the onboard computer, and the bipropellant thrusters. And to the delight of everyone, both the large manoeuvre thrusters and the smaller attitude control thrusters had worked perfectly. As for the anomalous yaw drift, since the adapter had been jettisoned for re-entry the thrusters could not be examined but the engineers concluded that the continuous force was caused by evaporation from the water boiler, and this design flaw could readily be rectified. In retrospect, an 'open ended' mission would have yielded more data on how the systems degraded, but it was a *new* spacecraft and, not unreasonably, the managers had erred on the side of caution.

Oddly, the aspect of this historic flight that attracted the greatest Press coverage was the corned beef sandwich. Grissom's favourite fayre, this had been bought the previous evening by Schirra at Wolfie's Deli in Cocoa Beach, which was a regular haunt for the astronauts at the Cape. He took the 'flight article' for this unofficial experiment (which was sanctioned by Slayton) to the suit-up van and surreptitiously passed it to Young, who put it into the knee-pocket of his suit. When a member of Congress criticised NASA for having "lost control of its astronauts", who had risked damaging the spacecraft with weightless crumbs, the House Appropriations Committee held an investigation, called in 44 of the agency's employees to testify and criticised James Webb, George Mueller, the Deputy Associate Administrator for Manned Space Flight, and Robert Gilruth, the Director of the Manned Spacecraft Center.[9] The other item of cargo carried by Gemini 3 was a 'Stars and Stripes' which was flown at the Manned Spacecraft Center for the remainder of the Program.

With this first test flight over, Schirra and Stafford switched their attention to their own mission (with Grissom and Young backing them up) which would attempt to make the first rendezvous and docking.

[9] In fact, Schirra had pulled a similar stunt for Gordon Cooper's Mercury mission when, while setting up the spacecraft, he had smuggled aboard a 'special ration kit' of a steak sandwich, a miniature of scotch and five cigarettes, although clearly in a pure oxygen environment the provision of a lighter was out of the question. Cooper, of course, refrained from availing himself of Schirra's kind offering.

Chapter 3

Stepping Out
Gemini 4

Something spectacular

When studies for spacewalking – or 'extravehicular activity' (EVA) as NASA decided to refer to it – started in 1962, there were no plans at that time to make EVAs on Apollo lunar missions,[1] but given the choice of Lunar Orbit Rendezvous it at least offered a contingency for transferring between vehicles in the event that a lunar module returning from the Moon was unable to dock with its mothership. And in the longer term, it was obvious that working outside a spacecraft would become necessary when NASA started to build a space station in Earth orbit.[2]

In January 1964, Gemini planners began to consider assigning the first EVA investigation to an early mission, but this was not a high priority and was contingent on the development of the necessary special apparatus and the operational procedures to use it. The David Clark Company of Worcester in Massachusetts, which supplied spacesuits, was asked to make one capable of withstanding the harsh space environment,[3] AiResearch in Los Angeles, which was making the spacecraft's Environmental Control System, was asked to supply a ventilator unit for the spacewalker to wear on his chest to regulate the flow of oxygen from an umbilical, and Harold Johnson of the Crew Systems Division of the Manned Spacecraft Center devised a Hand-Held Manoeuvring Unit (HHMU) which, by squirting a jet of cold gas, would enable a spacewalker to control his position and orientation.

The step-by-step plan called for an initial experiment in which an astronaut was simply to swing open the hatch and poke his head and shoulders above the casing in order to assess direct exposure to the space environment. After a few minutes, he was to regain his seat and close the hatch. The objective was to demonstrate that this could be done (or to identify any difficulties involved) and thereby clear the way for a later crew to don the special apparatus and attempt a full egress. There was one overriding constraint – only the Pilot would go out, the Command Pilot would remain strapped into his seat and look after the spacecraft. There was a significant risk involved

---

[1] A distinction is drawn here between in-space EVAs and lunar-surface EVAs. Apollo would obviously involve walking on the Moon, but in 1961 it was not expected that spacewalks would be made flying to and returning from the Moon.

[2] In October 1961, Emanuel Schnitzer at the Langley Research Center suggested using Apollo spacecraft and launch vehicles to assemble a laboratory in space - he called this 'Apollo X'. Over the years, this became first 'Apollo Extension', and then 'Post-Apollo', 'Apollo Applications' and ultimately 'Skylab'.

[3] This was designated the G4C suit.

because if the spacewalker were to become incapacitated, his commander, although fully-suited because the vehicle would be depressurised, would not be able to retrieve his colleague and draw his inert body into the seat, he would have no option but to unplug the spacewalker's umbilical and close the hatch.

At the Press conference in July 1964 to announce that Jim McDivitt and Ed White were to fly the Gemini 4 mission, which was at that time to be a 7-day test of the innovative fuel cell system, Deputy Program Manager Kenneth Kleinknecht raised the possibility of White opening the hatch but, to his surprise, the reporters let this pass without comment. In fact, behind the scenes White argued that he should don the chestpack and – after standing in the hatch to assess his reactions – egress to test the HHMU as a mobility aid, but the managers refused to commit to such an ambitious initial foray. In November 1964, while Gus Grissom and John Young were in training for Gemini 3, they voluntarily extended a test in a vacuum chamber to establish that a spacesuited astronaut could open and close the big, hinged hatch, and reported that it was difficult to close. Although this cleared the way for an EVA on one of the early missions, the special apparatus was still under development and so there was no guarantee that White would gain a full egress. On 12 March 1965, Robert Gilruth authorised another vacuum chamber test using Spacecraft 4, whose hatch had been modified to make it easier to close. When Alexei Leonov made a spacewalk a week later Gilruth and his deputy, George Low, backed White's proposal. NASA's medics – always fearful of human frailty – had initially argued that if a freely-floating spacewalker lost sight of his spacecraft he might become incapacitated by vertigo, but newsreel footage of Leonov floating with the Earth slowly drifting by gave no hint that he had suffered any such difficulty.[4] After Gemini 3's trouble-free flight on 23 March, Gilruth, Low, Dick Johnston of the Crew Systems Division and Warren North of Flight Crew Operations met to discuss the scope of a Gemini 4 EVA. When George Mueller, the Deputy Associate Administrator for Manned Space Flight,[5] paid a visit to Houston on 3 April he was lukewarm to attempting a full egress, on the basis that it would be better to stick to the step-by-step plan and have White merely open the hatch; he wanted to preclude criticism that NASA was 'reacting' to the Soviet Union's achievement. However, the special apparatus was now ready, and by the end of the month McDivitt and White were performing full-scale EVA simulations in the vacuum chamber and (for the first time) working with Flight Operations to develop procedures to firm up the flight plan – this was a significant step towards making White's EVA a reality, because until an item made it onto a flight plan it was not part of the mission. After witnessing a training exercise on 14 May, Associate Administrator Robert Seamans decided to recommend a full egress to James Webb and his deputy Hugh Dryden, but Dryden sided with Mueller. The *official* plan called

[4] The fact that Leonov's suit had 'ballooned', and he had been able to re-enter the airlock only after partially deflating his suit had not been reported. Even if this had been known, it would not have worried White because his suit was different to Leonov's and Gemini did not have an airlock. White would simply have to swing open the hatch and float out, and the vacuum chamber tests suggested that resuming his seat and closing the hatch ought to be straightforward.

[5] Mueller was recruited to supersede Brainerd Holmes in this role in 1963.

for a stand-up EVA on Gemini 5 (which was now to tackle the 7-day objective) and the first egress EVA to be made on Gemini 6 after its rendezvous and docking with an Agena target vehicle. On 19 May Charles Mathews informed Mueller that all the pieces were in place for an egress. Seamans briefed Webb and Dryden accordingly. On 25 May, Webb gave the go-ahead because spacewalking was a Program objective, the special apparatus was ready, and White was enthusiastic. Underlying Webb's endorsement of the plan was a recognition that the development of the Agena target vehicle was running so late that progress would not be able to be made towards the rendezvous objective until later in the year, and White's egress would be "something spectacular" to motivate the troops. The Gemini 4 Press Kit issued on 21 May had included 'Possible Extravehicular Activity', so the Press was told that if conditions permitted, White would give it a try.

There was a slight problem, however. The initial plan for a full-egress EVA on Gemini 6 had called for the spacewalk to be conducted in conjunction with the docked Agena target vehicle, but Gemini 4 did not have an Agena. Gordon Cooper had joked with Gus Grissom that if he sighted the upper stage of his Titan he should proceed to rendezvous. Gilruth and Low had overheard this remark and noted that 'station-keeping' with the spent stage would make an interesting experiment. McDivitt agreed to turn his spacecraft around immediately after separating from the upper stage, and fly alongside it. Since Gemini 4 would not have a radar with which to track the spent stage while in the Earth's shadow, the Martin Company added a pair of flashing lights. About an hour into the mission White would go out to assess the HHMU as a mobility aid, venture over to the stage and snap close-up pictures of it. Of course, photographing the Titan was make-work for White, but the exercise offered a good opportunity to evaluate the manoeuvrability of the spacecraft when flying in formation with another vehicle, as a rehearsal for drawing alongside an Agena.

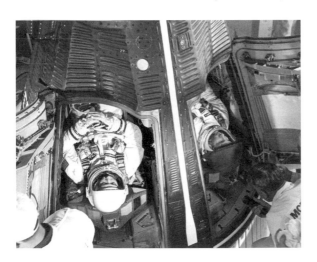

McDivitt and White ingressing their spacecraft.

Launch

McDivitt and White were awakened at 04:10 local time on Tuesday, 3 June 1965, for the now standard low-residue breakfast. At Pad 16 Joe Schmidt and Clyde Teague assisted them into their suits. On arriving at Pad 19 at 07:07 there was an hour and a half left on the clock. Having configured the spacecraft, the backup crew of Frank Borman and Jim Lovell went to join Al Shepard, serving as 'Stoney' CapCom in the blockhouse. All went smoothly until 08:25, when the hinged erector was to be lowered. After rotating 12 degrees it became stuck. It was raised back to vertical and operation restarted, but it stopped at the same point. Technicians returned to inspect it, and found that a plug had been incorrectly installed at a junction box, so they fitted it and withdrew, and the erector was lowered without incident. Despite this 76-minute hold at the T–35-minute point, the Titan was successfully dispatched at 10:16. Its ascent into the clear blue Florida sky was relayed 'live' to Europe (although in monochrome) by the newly commissioned Early Bird geostationary relay satellite stationed high above the Atlantic.

"We're on our way, buddy," McDivitt told White on the intercom as the launch vehicle rolled onto the correct azimuth and then began its pitch-over to head out over the Atlantic. "Boy, that Sun is bright!" he added as the early morning Sun shone in through the windows and illuminated the cabin so brightly that it was difficult to read the instruments. The ascent was a perfect ride. By SECO, the upper stage had inserted itself and the spacecraft into the planned elliptical orbit ranging between 88 and 153 nautical miles.

Gemini 4 lifts off on 3 June 1965.

With the Mission Operations Control Room at the Manned Spacecraft Center now the primary control facility, as soon as the Titan lifted off the focus of attention switched from the Cape to Houston. While the Press corps was well represented at the Cape, the prospect of an EVA on the second orbit prompted a strong news presence in Houston – indeed there were so many reporters registered that NASA had to lease a nearby office as an impromptu news room, and rig a 'squawk box' for the commentary by the Public Affairs Officer, Paul Haney.

The station-keeping experiment

After the second stage shut down, White threw the switch to detonate the pyrotechnics around the rim of the adapter to separate the spacecraft. Once free, McDivitt fired the aft-facing OAMS thrusters to move clear at 10 feet per second. After 30 seconds, he cancelled this motion and yawed around 180 degrees. As they passed over the Bermuda station at the far end of the Eastern Test Range, he informed Gus Grissom in Houston that they were 300 to 400 feet from the spent stage, which had evidently been disturbed by the pyro charges – because the 17-foot-long 10-foot-diameter cylinder was tumbling at a rate of 30 degrees per second. As the stage had a dry mass of 6,000 pounds, this instability meant that they would have to remain well clear of it during station-keeping. Another reason to keep their distance was that they did not want to contaminate the skin of the spacecraft with the propellant that was venting from the engine, whose large thrust chamber projected 10 feet from the rear of the cylindrical tankage. As this was the first time that anyone had seen a spent stage in orbit, the rotation and the efflux came as a surprise. Clearly, both factors would constrain White's activities, because if he managed to manoeuvre across to the stage his suit might get coated with toxic chemicals, and his umbilical might easily become tangled up. The EVA had been scheduled for early in the mission because the spacecraft did not have propellant for lengthy station-keeping, and because the doctors thought that it would be safer if he went out while he was fresh, rather than after his condition had deteriorated to the degree that he would be unable to deal with an emergency; in fact, the plan was for him to open the hatch soon after orbital sunrise as they passed over Hawaii on the second orbit.

McDivitt aimed his spacecraft's nose directly towards the stage and thrusted towards it at 5 feet per second. Surprisingly, it receded. He added another 3 feet per second, and was astonished to observe the range open further. Strange! Furthermore, after separating he had been out in front, but after manoeuvring he had drifted off to one side and above the stage.

"Ask him how far he is away from the launch vehicle," Chris Kraft, the Flight Director, prompted as the spacecraft flew into range of the Canary CapCom.

"I'm aligning the platform," McDivitt reported. In order to configure the platform of the inertial guidance system he had had to turn away from the stage, and therefore did not know how far away it was. A few moments later, he swung around. "I have the stage in sight, it's about directly below us at about four or five hundred feet." By now they were crossing the Sahara – which made a striking backdrop – but they had not yet unpacked a camera and so the opportunity to document the experiment was missed.

After crossing the terminator into darkness they kept track of the stage by its acquisition lights. Over the Indian Ocean, they were called by Ed Fendell at Carnarvon on the western coast of Australia, and McDivitt explained that he had not been able to return to the stage, whose lights he could still see. It was difficult to estimate the distance but he thought it was "probably around half a mile".

As they approached the United States, the Guaymas station in Mexico picked them up. "Is there any way of giving me a quick check on my orbit?" McDivitt asked.

"We're working on it."

Kraft called Guaymas on the ground circuit: "Tell him it will probably be after this pass over the States before we can get it to him." Radar tracking by Carnarvon had to be included to determine the parameters of the orbit.

"Gemini 4," Guaymas relayed, "we'll update you with the orbit calculations some time after the pass over the States."

McDivitt was dissatisfied. "We're going to have to get a resolution right away – do you want me to really make a major effort to close with this thing, or to save the fuel?" Whereas the range had initially opened, now that they were dropping towards perigee it was closing again and there was the prospect of a second opportunity to draw alongside the spent stage.

"I don't think it is worth it," Kraft opined. "Tell him that as far as we are concerned we want to save the fuel."

"Gemini 4, Flight advises that he'd like to save the fuel," Guaymas relayed. Advice from the Flight Director constituted an order. But then he added ambiguously, "You'll be advised over the Cape."

"I just can't wait 'till I get to the Cape," McDivitt insisted. If he was going to chase the stage he had better start now.

"Tell him to forget it," Kraft ordered.

"I guess we'll scrub it," Guaymas told McDivitt.

"Okay," McDivitt conceded. "Get me a tag on my orbit as quick as you can."

"We're working on it." In fact, as a result of its manoeuvring the spacecraft's orbit now had a perigee of 88 nautical miles and an apogee of 157 nautical miles.

A few minutes later, Gus Grissom in Houston explained the decision. "Jim, we estimate that you've used about 50 per cent of your OAMS capability at this time." The manoeuvres had used 160 out of the available total of 360 foot per second delta-V. As the OAMS was to make a 'fail safe' burn prior to the solid retro rockets firing, they did not have the propellant to spare to chase the spent stage. The later spacecraft that were to rendezvous with Agenas would have larger propellant tanks.

"I think we ought to knock it off, Gus," McDivitt agreed. "It's probably three or four miles away, and we just can't close up."

"Right, knock it off," Grissom reiterated, "no more rendezvous with the booster." If the station-keeping exercise had proved straightforward, some additional manoeuvres had been set for the fifth orbit, following White's extravehicular activity, but these were now deleted from the flight plan.

McDivitt had been caught out right from the start. When the upper stage of the booster shut down it was aligned along the velocity vector which, since the orbit was elliptical, was inclined slightly upward. In separating, he had made a prograde/up burn which produced an apogee about 90 degrees around from the insertion point and the Titan seemed to pull ahead and descend as the orbits diverged. When he tried to return by propelling himself towards his target with Gemini 4's nose angled down, he actually *increased* the divergence between the orbits. As no one had warned him otherwise, he had expected to be able to fly in formation with the spent stage using his 'seat of the pants' instincts as a pilot. The subtlety of orbital dynamics had taken him by surprise. As Andre Meyer, a trajectory specialist, sympathised, McDivitt "just didn't understand". However, this was progress in its own right because the point of the Program was *to learn*. "Frankly," Deke Slayton reflected, the station-keeping "wasn't an operation that had been too well thought out." Of the senior managers, Robert Seamans was the best placed to appreciate McDivitt's frustration, because prior to joining NASA he had worked on rendezvous concepts for the Air Force, in planning for a satellite interceptor known as SAINT. He had a team of specialists at the Langley Research Center report on what had just happened. When they concluded that the crew's training had been inadequate, training aids were developed as a priority to help the crew assigned to the first rendezvous mission. As a result of McDivitt's discovery that 'eye ball' flying did not work, "we got a whole lot smarter", pointed out Meyer. Gemini 4's frustration paved the way for the later missions on which rendezvousing, station-keeping and docking would be essential objectives.

White inside the spacecraft with his helmet on.

EVA preparations

While McDivitt's attention was on station-keeping with the spent stage, White checked his suit's integrity as a preliminary to the EVA. As they passed over Canary Island on the second orbit, he retrieved his special apparatus from its stowage lockers and began to work on the 40-item checklist. It took longer to strap the Ventilation Control Module (VCM) onto his chest harness and to unpack the umbilical and the HHMU than it had in training, but he was in no hurry. Although both men would be exposed to vacuum with the hatch open, only White was wearing the 21-layer G4C suit designed to protect against the 300° temperature difference between sunlight and shade. The G4C was essentially the G3C suit that McDivitt wore, fitted with an extra outer covering one-fifth of an inch thick comprising two layers of nylon for protection against micrometeoroids, seven layers of aluminised mylar for thermal insulation, another micrometeoroid layer and an outer nylon skin. His helmet was adapted to support a double external visor, the outer plexiglass visor of which had a gold film to reduce the transmittance by 88 per cent, with the inner polycarbonate visor filtering out the harmful ultraviolet sunlight and acting as a heat barrier to restrict the inward passage of solar infrared during daylight and heat loss through the faceplate in darkness. He was to don thick thermal gloves over the gloves of his suit to protect his fingers in the event of touching a part of the spacecraft's skin that was frozen. The VCM would draw oxygen from the 25-foot umbilical in such a way as to regulate the suit's pressure, and contained a tank with sufficient oxygen for a 10-minute ingress procedure if the umbilical failed. In addition to the oxygen hose, the umbilical incorporated wires for the communications system and a nylon safety tether.

Grissom called through the voice-relay station at Kano in Nigeria. "Go ahead with the attitudes you planned – that is BEF and 180 degrees roll." BEF meant 'blunt end forward'. "That should be best for photographic purposes." In writing the flight plan, everything that the astronauts would need to know had been worked out in advance.

"Roger," McDivitt acknowledged.

Grissom then called White: "Your best sun-angle for getting a picture of the spacecraft would be to move out forward."

That is, McDivitt was to fly backwards and inverted, and White was to position himself over the spacecraft's nose to take pictures looking towards the hatch while McDivitt took a picture of White with the Earth in the background.

"It's getting a bit crowded in here, Gus," the astronauts reported as they wrestled with the bulky EVA apparatus.

"I'll bet."

On establishing communication with Carnarvon, McDivitt told Fendell that White was running late. "Right now, he's trying to get his umbilical on. There's a lot to do, and we're having trouble keeping track." Once White was on the umbilical, he was to disconnect from the spacecraft's Environmental Control System. His final act prior to depressurising the air from the cabin would be to check his emergency apparatus. When Fendell relayed the 'Go' from Houston, McDivitt issued a warning. "Advise Houston that we're running a little late, and we might not be ready at Hawaii." The spacecraft's cabin was to be depressurised from its 5.2 psi norm slowly, so that if one of the suits was found to have a leak the cabin would be able to be repressurised rapidly. McDivitt doubted that they would be ready to open the hatch at the scheduled time without truncating this test, which he did not wish to do. There were three constraints. The first was that they must be in communication with Hawaii when the hatch was opened, so that Houston would know if something went awry. Secondly, the excursion had to be in daylight, and while overflying the United States for near-continuous communications across the continent. And White must ingress before they flew beyond the Eastern Test Range so that Houston would know that they were safely 'buttoned up' before they crossed the terminator.

Kraft instructed Fendell: "You tell him that if he doesn't make it on this pass, we'll take his evaluation and do it next pass."

Passing over the Canton Island relay station heading toward Hawaii, McDivitt reported that they were ready, but upon establishing contact with Hawaii a few minutes later he said they had decided to delay the EVA until the third orbit because White had become hot and sweaty. "I don't think we want to try it."

"Understand, next pass around," acknowledged Stu Davis, the Hawaii CapCom.

"Tell him we're happy with that," Kraft agreed. "Tell White to get back on the normal suit circuit."

"Flight advises, that you go back on normal suit circuit until next time around, then go back [to the umbilical] over Carnarvon [as before]."

As the spacecraft approached the California coast, Grissom called through the relay site at Point Arguello: "I think you did a smart thing, back there."

"It would have been impossible," McDivitt admitted. "We'd never have gotten the EVA done at all."

"Get him to describe his status inside the cockpit," Kraft prompted.

"How about describing the way the cockpit is laid out now," Grissom dutifully enquired, "with all of your gear out."

"When we finally called it quits, it was obvious that we weren't going to make it without really rushing, and I didn't want to do that," McDivitt said. "I've got the gun, the camera and the hatch fitting. He's got all the paraphernalia on right now, but is on the suit circuit. We're just about all set to go. Over Africa, we're going to go through the checklist again, and when we get to Carnarvon we'll be all set."

"Is Ed still there!?" Grissom asked as they flew over Texas.

"Yeah," McDivitt chuckled. "He just doesn't like to talk, I guess."

"Did you get pretty heated up, getting all that gear out?" Grissom asked White.

"I got pretty warm," admitted White.

"How do you feel now?"

"Fine."

They crossed Africa and flew on towards Australia on the third orbit. "I understand we have a 'Go' to start depressurisation, is that right?" McDivitt asked Fendell.

"That's affirmative."

The imminence of White's EVA had not been announced until the spacecraft was safely in orbit. When they heard on the radio that he was about to open the hatch, Pat Collins and Susan Borman were flying back to Houston after having watched the launch, and they were taken aback. "My God," Mrs Collins exclaimed, "he's getting out!"

Okay, I'm out!

As White opened the hatch's lock, the ratchet's action was impeded by a faulty spring, but because he had completely disassembled a hatch at McDonnell he was familiar with its mechanism and after diagnosing the fault he used brute force to overcome it. In the trials in the vacuum chamber he had found the heavy hatch difficult to open, and while he had expected it to open more easily in space he found it as he later reported, "as hard to push up in space as it had been on the ground." After swinging it wide, he mounted a guard on the sill to prevent his umbilical from becoming snagged.

Standing up in his seat, White installed a Maurer 16-mm movie camera facing forward in order to record his excursion. There was no indicator to show that the camera was operating, and trying to feel the vibration of its motor was impossible through the thermal glove, so he took one off and checked the camera three times to verify that it was properly configured. "I wanted to make sure that I didn't leave the lens cap on," he quipped later. "I knew I might as well not come back, if I did!" Deciding that he did not have time to put the glove back on, he tossed it back into the spacecraft. The movie camera was loaded with 100 feet of colour film and set to expose six frames per second, so it would have an endurance of about 15 minutes. Because the EVA was a late addition to the flight plan, the camera had been loaded onboard only a few days before liftoff. McDivitt was to aim a 70-mm Hasselblad through his window and snap pictures of White as and when an opportunity presented itself.

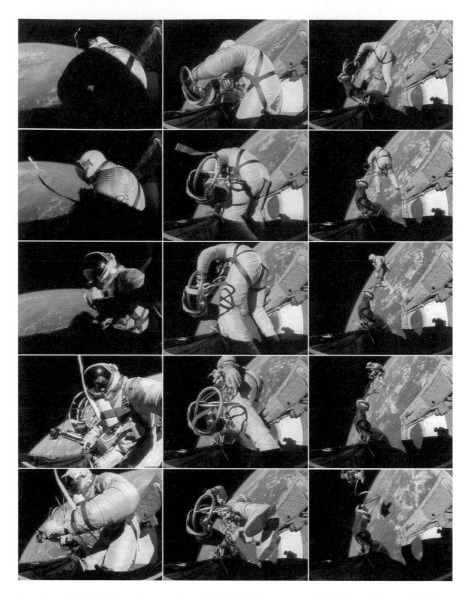

A series of stills from the movie of Ed White's spacewalk. (above & opposite)

"He's standing in the seat," McDivitt told Davis when communications with Hawaii were established.

In Houston, Dr Charles Berry, the Flight Surgeon, told Kraft that White's biomedical telemetry was satisfactory. "Let's go."

"Tell him we're ready to have him get out when he's ready," Kraft told Hawaii.

"We just had word from Houston," Davis relayed. "And we're ready to have you get out whenever you're ready".

"We've got our 'Go' now, is that right?" queried McDivitt.

"Affirmative."

"We've still got a little work, right here."

"Roger, understand."

White assembled the two-part HHMU 'zip gun' by fitting the vent tubes onto the

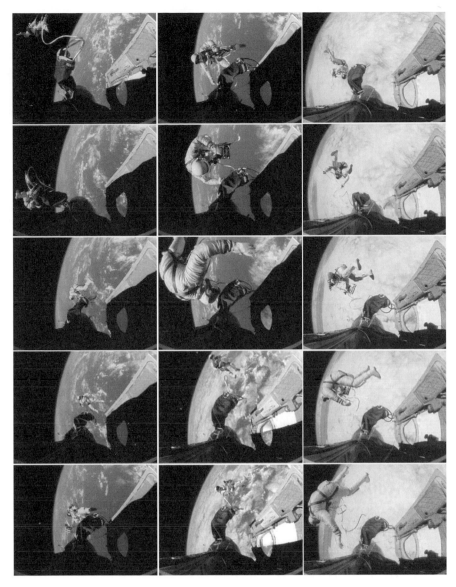

hand unit, which integrated the trigger and a pair of canisters of the same type as used to supply emergency oxygen to the ejection seats. It had been decided to use oxygen as the propellant so that if the gas leaked into the cabin it would not compromise the all-oxygen environment. Having assembled the gun, White tethered it to his right wrist so that he would not lose it if he released his grip.

McDivitt set the communication system to VOX, so that the ground would be able to listen to White on the intercom circuit. By the time that White was ready to exit, they had left Hawaii behind.

"I'm separating from the spacecraft," White said, beginning his running commentary to anyone who was listening and to the onboard tape recorder. With that, he gently pushed off. "Okay, my feet are out – I think I'm dragging a little bit, but I don't want to fire the gun yet. Okay, I'm out!" He used a squirt of gas from the HHMU to move out over the spacecraft's nose into the field of view of the movie camera so that it could record his mobility and stability evaluation. To make a 'pure' translation motion, he had to aim the gun in order that the impulse would be directed through his centre of mass. If he did not line it up accurately, the offset force would impart a rotation as well as a translation. There was a thumb-selector on the trigger to chose whether to squirt through a centrally mounted forward-firing vent, or through aft-firing vents at the ends of the side arms. There was no specific test sequence, he simply had to test as many control inputs as he could with the small amount of gas available. "Okay, I put a little roll in there," he reported as he began a slow tumble. "I took it out." He had cancelled the unwanted motion. "I'm rolling to the right now under my own influence."

As White looked back towards the spacecraft, he saw something float out of the hatch. "It looks like a thermal glove, Jim," he reported. The glove that he had tossed back into the spacecraft had drifted out and away, its departure recorded by the movie camera. McDivitt was aware of a 'flow'. The materials inside the cabin were out-gassing and his pressure suit was evidently leaking a small amount of oxygen, and this flow had swept up the glove.

"Alright, now I'm coming above the spacecraft." White translated up. "I'm coming back down now." He came back down. "I'm under my own control."

"You're right in front, Ed," McDivitt advised. "You look beautiful."

After halting, White turned to face the spacecraft. "I feel like a million dollars!" he said delightedly. "I'm coming back to you." He started back towards the spacecraft. "The gun works real good, Jim."

As White crossed over the spacecraft, he tugged on the umbilical to see if he could use it to control his movement, but it set him tumbling and before he could regain control using the gun he had drifted all the way back over the adapter module and out of the movie camera's field of view. After stabilising himself with the gun, he moved forward above the spacecraft's centreline and once again brought himself to a halt out in front of the nose.

Since leaving Hawaii behind, the spacecraft had been out of radio contact. Houston was eagerly waiting for Guaymas to pick them up, but while the spacecraft's telemetry had been received there was nothing on the voice circuit. Grissom had to call several times to attract McDivitt's attention.

"Who's calling Gemini 4?" McDivitt enquired.

"Has he egressed?" Grissom asked, getting straight to the point.

McDivitt realised that the VOX had not been working, so he reset it. "He's out, Gus!"

"That's great," Grissom agreed.

Meanwhile, White realised that he had run out of gas for his gun. "I think I've exhausted my air now. I've had very good control with it. I just needed more air." The trial had taken about 4 minutes. Having exhausted his manoeuvring gun's supply of gas, he had to resort to trying to control his movements by swinging the umbilical but, as his earlier experiment had hinted, this was not very efficient. As White slowly tumbled, he was finally able to relax and admire the view. Inside the cabin, his field of view had been limited by the small half-moon-shaped window, but now the world filled his visor. The medics had fretted that the sense of being 'out there' might induce vertigo, especially when he could not see his ship, but he was delighted. "This is the greatest experience, it's just tremendous. Right now, I'm standing on my head and I'm looking down. It looks like we're coming up on the coast of California. I'm in a slow roll to the right. There's absolutely no disorientation associated with it." The clarity of the view was incredible. "I can see much greater detail than I can from an aircraft flying at 40,000 feet."

As White tugged on his umbilical to try to cancel his rotation, the umbilical transmitted the force to the spacecraft, which reacted in response. "One thing about it," noted McDivitt as he fired the thrusters to hold the craft stable, "when Ed starts whipping around that thing, it sure makes the spacecraft hard to control."

One of the few pictures snapped by White while he was outside using the camera on his 'zip gun'.

White had a 35-mm Conterax camera on top of the HHMU and he fired off a few frames as his slow tumble presented opportunities. Meanwhile, as White floated in front of his window, McDivitt was taking pictures. "Hey, Ed, smile!"

"I'm looking right down your gun barrel, huh?"

"Let me take a close-up picture of you," McDivitt said, as White continued to approach the spacecraft. They had wanted to use the call sign '*American Eagle*' but James Webb had banned the naming of spacecraft, so they had decided instead to wear the flag – and in doing so started a tradition. With the 'Stars and Stripes' prominent on White's shoulder, the picture became an icon of America in space.

As White made contact with the spacecraft, his free hand rubbed up against McDivitt's window.

"You smeared up my window shield, you dirty dog!"

"Yes?"

"It looks like there's a coating on the outside, and you rubbed it off – that's apparently what you've done." In fact, as on Gemini 3, the left window had mysteriously picked up a thin smear during the ascent through the atmosphere.[6] Once White was safely back inside, Houston asked him to store the thermal glove in a plastic bag so that this coating could be analysed later.

"It looks like we're about over Texas," McDivitt mused. "As a matter of fact, that looks like Houston below."

"Yes, that's Galveston Bay right there," White agreed.

"I'm going to Push-to-Talk and see what the Flight Director has got to say," McDivitt told White. "Hey, Gus, this is Jim. Got any message for us?"

[6] It was concluded that this film was picked up during staging, when the fireball washed over the nose of the spacecraft.

As White floated outside his window, McDivitt snapped this striking
picture capturing the Stars and Stripes shoulder patch.

In fact, Grissom had been trying to make contact for several minutes without attracting McDivitt's attention.

Kraft pushed his communications switch: "The Flight director says 'Get back in'!"

"Get back in," Grissom echoed.

"Okay," McDivitt acknowledged.

"Where are we over now, Jim?" Being on the intercom, White had not heard Houston's call.

"I don't know," McDivitt dismissed, because the time for sightseeing was over. "They want you to come back in now."

"Back in?"

"Back in," McDivitt confirmed.

"We have been trying to talk to you for a while," pointed out Grissom, but there was no indication that McDivitt had heard him.

"I'm coming," White acknowledged. He pulled on his umbilical to draw himself towards the spacecraft and planted his feet on the adapter to stabilise himself, before pulling himself to the open hatch.

"You've got about four minutes to Bermuda LOS," Grissom warned, again without any indication that he had been heard. After Bermuda there would be a gap to Canary and Kraft wanted the spacecraft buttoned up. Certainly, he did not want White floundering in the dark, and because the EVA had been put off for an orbit the Earth's shadow was about 20 degrees closer than it otherwise would have been, and it was not evident that the crew realised this.

White took a final look down at Cuba and Puerto Rico and then swung himself around to ease his feet in through the hatch onto his seat. After retrieving the movie camera, he gave it to McDivitt to stow.

"Come on, let's get you back in here before it gets dark," McDivitt urged, and he started to draw in the 25 feet of umbilical, stuffing it into his footwell out of the way.

"It's the saddest moment of my life," White reflected.

"Well, you're going to find it sadder when we have to come down with this whole thing," promised McDivitt, referring to when they would de-orbit and return to Earth.

"I'm coming."

McDivitt switched to the radio circuit. "Any messages for me, Houston."

"Are you getting him back in?" Grissom asked.

"His legs are down below the instrument panel."

"You're going to have Bermuda LOS in about 20 seconds."

"He's having some trouble getting back in the spacecraft."

"Are your cabin lights up bright?" Grissom prompted. "In case you hit darkness."

McDivitt's reply faded out as the spacecraft flew beyond the Bermuda station's range.

"What did he say, Gus?" Kraft asked.

"He said that he was busy, and would rather not talk to us," Grissom replied, offering an astronaut's interpretation of the situation.

White was having difficulty bending his legs and waist in the inflated suit to ease down into his seat, and his view out of the lower part of his visor was limited, but he managed to force his knees down under the instrument panel and use the resulting leverage to force his bottom down onto the seat. Although he was able to pull the hatch shut without too much difficulty, the fouled ratchet made the mechanism difficult to lock. Applying torque merely hoisted White from his seat, so McDivitt held him down while he engaged the lock. It took several minutes, by which time White was drenched in sweat. His exertion overwhelmed the suit's environmental system, and his faceplate misted up. His heart rate had hovered at 150 beats per minute during the 36 minutes that the hatch had been open, but shot up to 180 as he struggled to lock it. The flight plan called for him to switch back to the standard air hose and then swing the hatch open again for a few minutes to jettison the VCM, umbilical and HHMU, but McDivitt cancelled the reopening – they would simply have to find space for the stuff in the spacecraft.

The headline in the local newspaper the next day was '*Ah, Ed. Please Get Back In*

The Spacecraft' and the story interpreted Houston's repeated 'get back in' calls as meaning that White – a military officer – had refused an order. Some medics suggested that he might have been suffering from a state of euphoria, which was what they had warned against. What the newsmen had not appreciated as they heard Grissom repeatedly call the spacecraft, was that McDivitt, on the intercom, was only checking the radio periodically and as soon as he heard the instruction he ordered White back in, and White did as ordered. Such was the gulf between those learning to fly in space and those reporting on the endeavours.

The lesson learned by NASA from this experiment was that although the hatch required attention, White had not suffered any disorientation and had been able to manoeuvre using the HHMU. Overall, it looked as if working outside the spacecraft would be straightforward. Also, as NASA promptly pointed out, whereas Leonov had been out for 10 minutes, White had been out for 21 minutes and, at last, America held an endurance record in space.

Clocking up time

Having been told of McDivitt's request to be left alone, Ascension monitored the spacecraft's telemetry without attempting to communicate.

"Hello, Carnarvon," McDivitt replied when Fendell announced that he was standing by, "it's nice to have someone to talk to again." As expected, he reported that the cabin was up to full pressure and the hatch seal was airtight, and then he added that he had decided not to reopen the hatch to jettison the EVA apparatus. He did not explain why. The Public Affairs Officer in Houston speculated that White wanted to keep the apparatus "as a souvenir". On contacting Hawaii, McDivitt explained that he had taken this decision because the hatch had been difficult to close.

Communications were 'noisy' on their next pass across the US, but Dr Berry debriefed each man. When he asked White whether he had suffered any symptoms of disorientation, White was emphatic that he had not – it had been a pleasurable experience. Of hearing of White's exertion closing the hatch, Berry urged him to try to get at least four hours sleep as soon as possible.

Three teams of flight controllers had been trained to provide 24-hour coverage during the four-day mission. There would be a Press conference after each change of shift. Chris Kraft, who was serving as Mission Director as well as lead Flight Director, was to take the 'active' part of the day – having overseen the launch, station-keeping and EVA he now handed over to Gene Kranz. Formerly Kraft's Procedures Officer for Mercury, Kranz would monitor the spacecraft's systems during the less active periods. John Hodge had taken the 'slack' part of Gordon Cooper's 34-hour Mercury flight. He would do the same for Gemini 4, and firm up the flight plan for Kraft's next shift. In addition, Glynn Lunney, a Flight Dynamics Officer, sat in for on-the-job training to pick up the load later on. Clearly, with flight control facing a learning curve these arrangements were an experiment in their own right. Whenever feasible the shift changes would be made between communication sessions, with the members of the new team arriving 30 minutes early in order to be briefed by their retiring colleagues before taking over.

With Gemini 4 safely 'buttoned up', Kranz revised the flight plan to take account of the facts that McDivitt had used much more propellant than expected in trying to station-keep with the Titan stage and the delay of the EVA by one orbit. Since half of the propellant had been used, it was imperative that they conserve that which remained. As they had failed to stay with the stage, the follow-on manoeuvres which had been planned were deleted. It had originally been hoped to fly the first endurance mission with the fuel cell system providing sufficient power to enable the spacecraft to remain powered-up for most of the time, but the development of this innovative technology was running late and Gemini 4 was running on a set of batteries which, although they could readily sustain a two-day mission, would require the spacecraft to spend much of its four-day mission powered down. Although some of the experiments assigned would require the spacecraft to adopt a specific orientation, it was decided to let the spacecraft drift in attitude in order to further reduce the use of propellant.

Two shots of McDivitt inside the spacecraft, with and without his helmet.

Meanwhile, the astronauts had their first space meal. This was a mixture of munchies (a variety of items compressed into bite-sized lumps) and bags of freeze-dried and dehydrated items which would be reconstituted by water delivered by a squirt gun. The food packs were stored in a locker over McDivitt's left shoulder, and were linked together in sequence so that all he needed to do when meal time came around was to pull out the next pack.

As this was the first real test of the new spacecraft, it had been decided that one member of the crew must be alert at all times in case of a malfunction. Although White was exhausted by his exertions, it proved difficult to sleep because the flight plan called for them to keep their helmets on and the volume of the headset had been deliberately arranged not to be able to fall completely silent, so there was a constant chatter as a succession of stations announced that they were standing by and McDivitt reported in. After lightly snoozing for four hours, White was awakened. An hour later, McDivitt took his turn. The plan was for them to alternate in this way through the mission. One frustration was that when something had to be unstowed, it turned into a major exercise because the few lockers were inconveniently located, narrow and deep and if the item sought was buried, the items removed to access it floated around in the cabin, and had to be retrieved and stuffed back into the locker – it was impractical to try to pick out something on a whim.

White with his helmet off.

As the spacecraft drifted, slowly tumbling, the astronaut on duty would stand by with a camera in his window to snap pictures of weather features and interesting-looking terrain – during his Mercury mission Gordon Cooper had astounded people by reporting that he was able to see smoke trails from trains and ships at sea and this was the scientific community's first opportunity for significant orbital photography. McDivitt and White confirmed that it was possible to see airfields, highways and other fine details. In addition, sensors recorded the radiation both outside and inside the craft. In truth, however, after the excitement of the EVA the rest of the flight was rather tedious. It was an endurance test, firstly to establish the reliability of the spacecraft, but also of the ability of the human body to survive four days in space,[7] and so the main activity was taking and reporting temperature and blood pressure. One point of concern for the medics was that McDivitt reported an eye irritation, a dryness of his throat and a hoarse voice, suspected to be due to dust from the lithium hydroxide canister in the Environmental Control System that removed carbon dioxide from the cabin's air – although if that was the case it was a mystery why White showed no such symptoms.

One welcome break from the routine on the second day was a call from their wives. As a surprise, Kraft invited Pat McDivitt and Pat White into the control room to chat with their husbands. When asked what he was doing, Jim McDivitt replied, "All I can do is sleep and look out the window!" When asked if he could move around, he laughed and made an indirect reference to the 25-foot umbilical, "Ed is cluttering up the place." Mrs White told her husband proudly, "It looked like you were having a wonderful time yesterday", and he agreed that he had had "quite a time". A few hours later, Stu Davis in Hawaii congratulated them on breaking Gordon Cooper's American endurance record of 34 hours and 20 minutes. Now they were pioneering. When his wife called again the next day, White said that it was nice to hear a female voice after having listened to the gruff Grissom for seven hours. A few minutes later, their Titan stage re-entered the atmosphere and burned up over the Atlantic. With their oxygen, propellant and batteries all looking good, they were given the go-ahead to attempt the full four days envisaged by the flight plan. By

[7] This had already been established, of course, by the Soviets when Valeri Bykovsky flew for a few minutes short of five days on Vostok 5 in June 1963.

Mrs. Patricia McDivitt (left) and Mrs. Patricia White in the MOCR talk to Gemini 4.

T+58 hours, they exceeded the accumulated time in space for all of their American predecessors. Although powered down, the spacecraft was healthy. When a reporter asked at the end of the third day whether there was any intention to extend the flight, Kraft pointed out that although there were sufficient consumables for another day, there would be no extension. Back in 1964, the idea had been for the first 'endurance mission' to last a week, but the medics had argued for a progressive strategy, starting by 'doubling up' Gordon Cooper's flight, so this mission had been set at four days. The first malfunction occurred at T+76 hours. From time to time, McDivitt had been powering up the computer to enable its data to be updated by the ground. It had become routine. Now he reported that the on/off switch had failed, with the result he could not start it. After several hours characterising the problem, Kraft decided to forget the computer, and have them execute a "Mercury-type" return in which they would set the capsule rolling for a ballistic re-entry rather than controlling their attitude to use the capsule's aerodynamic lift to steer a specific trajectory.

They started stowing items for the return to Earth. Although White's EVA chestpack went back into its slot in his footwell, he had to roll up the 25-foot umbilical and stuff it in between his thighs and the seat casing out of the way. After the pre-retrofire checklist, Stu Davis in Hawaii counted them down to the 'fail safe' OAMS burn, which lasted 2 minutes and 41 seconds and used most of the remaining propellant. They jettisoned the equipment section of the adapter module shortly before making contact with the Guaymas relay station. McDivitt triggered the retrofire sequence 1 second late, and so they came in slightly 'long'. Houston monitored the ripple-firing of the four retro-rockets and the jettisoning of the retro section. As the capsule entered the atmosphere, McDivitt used the RCS thrusters on its nose to set it rolling at 15 degrees per second to cancel out the lift that its shape would otherwise generate. One result of a steep descent was a doubling of the deceleration load to 8 g, which McDivitt joked would be too much a man of his age. Looking out of the window during re-entry, he saw their retro section burn up. After the drogue chute and the main opened on time, mindful of what had happened to Gemini 3, just before transitioning to the two-point suspension they raised their arms to prevent their helmets striking the window frames. The capsule splashed down very heavily into the Atlantic some 530 nautical miles southwest of Bermuda.

The prime recovery ship, the aircraft carrier *USS Wasp*, was 36 nautical miles away, so once he was assured that the swimmers had fitted the flotation collar McDivitt requested a helicopter pickup. Watching the 'live' television of the astronauts, still in their suits (minus helmets and gloves) and with four days' of chin growth, stand in the helicopter's hatch and salute prior to stepping down onto the red carpet on the flight deck, Kraft lit his traditional end-of-mission cigar and shook hands with Gilruth. "We're feeling a bit tired," McDivitt said, "and we'd like something to eat." As they awaited *real* food, they received a congratulatory telephone call from President Johnson. After the post-flight debriefing back in Houston, NASA sent McDivitt and White to the Paris Air Show, where, by pure good luck, they met Yuri Gagarin, marking the first time that spacefarers from the competing countries had met.

White (left) and McDivitt after stepping down from the recovery helicopter onto the deck of the *USS Wasp*.

The medical team had been astonished to see White dancing with delight as he made his way along the carpet – not only did he not faint as the blood drained to his legs after so long in weightlessness, he did not even appear to be dizzy. While the post-flight tests revealed a surprising reduction in blood plasma volume (that is, cardiovascular capacity) in both men, Berry was delighted with the mission. "It was far, far better than anything we could have expected. We have knocked down an awful lot of 'straw men'. We had been told we would have unconscious astronauts after four days in weightlessness; well, they're not! We were told that the astronaut would experience vertigo when he stepped out of the spacecraft; we hit that one on the head." On observing McDivitt and White's rapid recovery he agreed to a further 'doubling up', and so the week-long flight of Gemini 5 was increased to 8 days, and Borman and Lovell were rotated to Gemini 7 to push for the Program's 14-day objective.[8] Although White was rotated into a backup role in this effort, Deke Slayton pulled McDivitt from the Gemini Program and assigned him to Apollo, to start tracking the development of the lunar module as a preliminary to taking command of the crew that would test it in Earth orbit.

[8] In effect, the Program's endurance objective was tackled by Gemini 4, 5 and 7, while the rendezvous theme was pursued independently by other missions.

White (left) and McDivitt below-deck on *USS Wasp*
taking a call from President Lyndon Johnson.

On tour after their mission Jim McDivitt and Ed White
meet Yuri Gagarin (left) at the Paris Air Show.

Chapter 4

'8 Days Or Bust'
Gemini 5

A challenging mission

Jim McDivitt's aborted effort to station-keep with his spent Titan stage had shown that flying a spacecraft was not like flying an aircraft. A radar and a computer would be required to conduct an orbital rendezvous. Spacecraft 5 was the first vehicle to roll of the production line with the radar installed. As spaceborne radar tracking was new, it had been decided that one of the early missions should release a Radar Evaluation Pod (REP) to test the ability of the radar to determine the range, range-rate and bearing angle to the target. The radar was expected to be capable of 'locking on' at a range of 200 nautical miles. It would feed data to the onboard computer (which was itself a real state-of-the-art gizmo) to compute the orbital manoeuvres for the rendezvous, so this experiment was also to determine how well the radar worked with the computer. In addition to the transponder, the 75-pound pod had been fitted with acquisition lights to enable it to be tracked visually during orbital night. But as the pod's battery would last only six hours the experiments would have to be conducted soon after its deployment.

In contrast to a chemical storage battery which releases *stored* energy, a fuel cell system converts the energy from combining its hydrogen and oxygen into electricity by sustaining a chemical process in which water is produced as a by-product. The process is the inverse of electrolysis in that instead of using electricity to break water molecules into their constituent atoms, it creates water from such atoms and liberates electrical energy, which is drawn off to power the spacecraft. The water would be fed into a tank, gaseous waste would be vented to space, and the heat produced would be dissipated by a cooling system. Fuel cells were to be used for Apollo because, for a given electrical load, a fuel cell system is much lighter than a storage battery.[1] The operating life of a fuel cell is determined by the supply of reactants, so for a longer mission it was necessary only to install larger tanks. A key aspect of the Gemini Program was to perfect this technology for Apollo. On Gemini, the fuel cell system and the tanks for the cryogenically chilled reactants were contained in the equipment section of the adapter module, and hence could be readily optimised for different

---

[1] A storage battery could be recharged using solar transducers, but Apollo was not to carry solar panels. The Soviet Union opted for solar panels to give its Soyuz long-duration capability.

missions. The fuel cells held out great promise, but managing fluids in weightlessness was a major 'unknown' factor in the design.

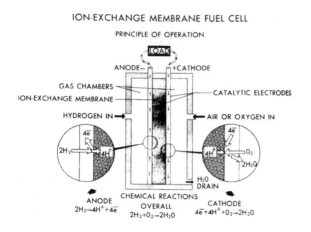

A schematic of the internal structure of a fuel cell power system.

When Charles Mathews took over as the Gemini Program Manager in March 1963, the plan was for the first manned mission to eject the REP in order to test the spaceborne radar and to fly for at least the 34 hours planned for Gordon Cooper's final Mercury mission that May. In April, however, Mathews revised the schedule and reduced the first flight to three orbits. As this would be insufficient to conduct the radar test, in July the radar and the REP were re-assigned to the next mission, on which there would be plenty of time as this was to be a seven-day test of the fuel cells. In October, with the development of the Agena target vehicle running late, Deke Slayton re-assigned the rendezvous from the third to the fourth mission. And with the development of the fuel cells also falling behind, Mathews accepted medical advice and split the early endurance target between the second and third missions, with one flying for four days – which was the most that could be attained by frugal use of batteries – and the other testing the fuel cells for seven days and performing the radar test. Given McDivitt and White's rapid recovery, the Gemini 5 objective was increased to eight days to match a circumlunar mission, because one option under consideration, if the Soviet Union looked like winning in the race to the Moon, was to send an early Apollo mothership on a 'free-return' trajectory on a loop around the Moon, so that NASA could claim to have been first to venture into 'cislunar' space. Initially, all Gemini spacecraft were to have used fuel cells, but protracted development problems prompted Mathews to decide in December to modify Spacecraft 3 to run off batteries. In early 1964 he ordered Spacecraft 4 similarly modified. As a contingency, in July 1964 a battery system was ordered for Spacecraft 5 but because a battery system capable of supporting such a long flight would be so heavy that it would preclude carrying the experiments that were to keep the crew occupied, to use it would impose a drastic penalty on the mission. In light of the fact that the output of a fuel cell falls as the load placed on it increases, on 14 August Mathews instructed McDonnell to develop a 'combined power system' for Spacecraft 5 in which batteries would support the peak loads and supplement the fuel cells which would sustain the spacecraft at other times, although this would still require deleting the experiments. On 6 November it was decided to run Gemini 6 off batteries, but as

Wally Schirra wanted a 24-hour mission uncluttered with experiments this was not a major problem. However, following progress with 'the high-load issue', on 18 December the battery-augmentation order for Spacecraft 5 was rescinded. The basic qualification test programme for the fuel cell system was completed in May 1965, just in time for the mission, so it was a close run thing.

Gordon Cooper and Pete Conrad were not officially announced as the crew for Gemini 5 until 8 February 1965 (just after the second unmanned test) with a view to launching in July following Gemini 3 in March and Gemini 4 in May, but the unprecedented pace produced a bottleneck in crew training and the schedule slipped, with Gemini 5 being postponed to mid-August.

The Gemini 5 crew had been provided with G4C suits when it was thought that Conrad would make the first full egress EVA, but when Ed White enhanced the basic hatch-opening test by floating out, and it was realised that nothing more could be achieved using the facilities available, the spacewalk was deleted from Gemini 5.

The 'Cooper Patch'.

Given that they would not be opening the hatch, Cooper sought permission to doff their suits once safely in orbit, but Mathews refused because even if this could be achieved there was no way that they would be able to rapidly suit-up again in an emergency. The increase from seven to eight days did little to improve Cooper's mood.

As the naming of spacecraft had been prohibited, Cooper and Conrad decided to create a 'mission patch', and in so doing started a trend that (together with wearing the flag) became standard practice.[2] If he had been able to name the spacecraft, Cooper would have selected '*Conestoga*' for the type of covered wagon that was used by the settlers who had opened up America, so as to capture the pioneering spirit of the flight. The patch featured a Conestoga wagon with '*8 Days or Bust*' on its canvas top. On 14 August James Webb accepted what he referred to as the "Cooper patch" providing the motto was elided, as its inclusion would only tempt the Press to deride the mission as a 'bust' in the event of it not flying the full duration. Cooper had a canvas rip-off strip sewn over the patch so that it could be exposed upon their triumphant return.

A rough ascent!

On 19 August Everett Christensen took over from Chris Kraft as Mission Director, and immediately had to scrub that day's countdown. Initially, there was a difficulty loading the 22 pounds of cryogenic hydrogen reactant for the fuel cells, and the clock

---

[2] There are patches for earlier missions, but these were designed retrospectively for commercial purposes.

Launch preparations: Gordon Cooper chats with Wally Schirra at breakfast (top-left);
Pete Conrad in the suit-up trailer on Pad 16 (lower left);
Cooper leads Conrad out to the transfer van (top right);
and Cooper walks up the ramp to the elevator on Pad 19 (lower right).

was held at T–5 hours until this was resolved. When the count was resumed, Neil Armstrong and Elliot See, the backup crew, set up the spacecraft while Cooper and Conrad breakfasted with Slayton, Wally Schirra and Tom Stafford. At T–100 minutes, Cooper and Conrad were sealed in the spacecraft. When a problem developed with the first stage of the Titan at T–30 minutes, a technician had to manually pump nitrogen into a 'stand pipe' in the propellant feed system after the pump failed to activate automatically. At T–25 minutes the erector began its slow rotation down. A problem with the telemetry system prompted a hold at T–10 minutes. As the delay extended, stormy weather down the coast migrated north. With the meteorologist warning of "a good possibility" of thunder, the erector was swung up to protect the launch vehicle. Conrad told Rusty Schweickart, the 'Stoney' CapCom in the blockhouse, that there were rain drops on his window, and jokingly asked for the windshield wipers to be switched on. "Let's hang on and try to go today," Cooper implored as lightning moved north towards the Cape. But the continuing telemetry problem prompted Christensen to scrub the launch. "You promised us a launch today, not a wet mock-up," Cooper rebuked the meteorologist.

Gemini 5 lifts off on 21 August 1965. Note the blockhouses inshore of the pads of Missile Row.

Two days later, they were back and this time the count ran smoothly down to launch at 09:00 on Saturday, 21 August.

"We're on our way," Cooper announced.

The Titan rolled to align its inertial guidance system on the designed azimuth and started to pitch over. "You're looking good," called Jim McDivitt, the CapCom in Houston.

"It feels mighty good," Cooper confirmed. "It's been a long time getting back!"

"You're 'Go' for staging," they were advised.

For 13 seconds towards the end of its burn, the first stage suffered a severe longitudinal oscillation known as 'pogo' (so-called on account of its effect being similar to riding a pogo stick) due to propellant sloshing in the tanks. The 10 cycles per second oscillation gave the crew an unpleasant ride, with the peak 0.38 g load exceeding the 0.25 g specified maximum. This oscillation was superimposed on the 3.3 g acceleration load at that phase of the ascent. The pogo came as a nasty surprise, as such oscillations had supposedly been 'designed out' when the Titan II missile was modified for use as the Gemini Launch Vehicle. Fortuitously, for the first time the stage was retrieved from the Atlantic. An investigation found that the 'fix' to overcome pogo – the oxidiser 'stand pipe' that had been manually charged when the system had failed to charge automatically – had received only 10 per cent of the volume of nitrogen required, which was why the effect recurred. The pogo faded a few seconds before the first stage shut down. The staging was nominal. To sustain the 'long' mission, the second stage flew a steep trajectory to produce the higher than usual apogee of 185 nautical miles.

For the first time, the first stage of a Gemini-Titan booster was recovered from the Atlantic. It had given its crew a rough 'pogo' ride.

Power crisis

As Gemini 5 set off across the Atlantic, the small batteries which had handled the high electrical load during the ascent were isolated as Cooper and Conrad worked through the post-insertion checklist. With luck, the batteries would not be required again until re-entry.

When they were picked up by Charles Lewis, the CapCom at Carnarvon on the western coast of Australia, Conrad told him that they could see the lights of the remarkably isolated city of Perth. Both men had a fondness for Australia. Cooper had served as the CapCom at Carnarvon for John Glenn's historic orbital mission, and Conrad had done so for Gemini 3. "We're right on the flight plan," Conrad assured. They were aligned SEF and ready to make the 10 foot per second perigee-raising burn on achieving apogee, so as to reduce the rate of decay of their orbit. After the burn, as they dipped towards his horizon, Lewis passed up an instruction from Houston concerning the fuel cells. The two tanks containing the cryogenic hydrogen and oxygen reactants each contained a heater that was to be operated briefly from time to time to maintain the pressure. These heaters could be set in 'AUTO' and triggered by sensors, or be operated manually. Lewis told them that they were to operate the heater in the hydrogen tank manually and regulate the pressure on the gauge on the console between 220 and 330 psi. Conrad noted that it was currently 250 psi. With that, they flew out across the Pacific unpacking a variety of apparatus that they would need for their early experiments.

As they crossed Texas, Conrad remarked to Cooper, expressing his delight with the fuel cells: "These fuel cells are four-oh, buddy, right down the line."

"Just like advertised," Cooper agreed. As they were the technological 'weak link' on his mission, he intended to run them at the lowest permitted pressure in order to stretch them to the full eight days.

Heading across the Atlantic for the second time, they took measurements for one of the biomedical experiments. On contacting Canary Island, they told Keith Kundel that the radar was in standby for the REP test. The fuel cells were a technological marvel, but they were high maintenance, as they had to be periodically purged to flush out the inert gases that did not participate in the reaction and other impurities. There were two 'sides' to the cell, with the oxygen reacting with the hydrogen via a solid platinum ion-exchange membrane which formed the divider, and the two sides were to be purged separately. At Canary's prompting, Conrad purged the hydrogen for 15 seconds.

Ten minutes after leaving Canary, passing over Africa towards the evening terminator, Cooper yawed 90 degrees, threw the switch to eject the REP from its housing in the rear of the adapter, and then yawed back around to aim the radar in the nose of his spacecraft at it. The radar locked on immediately, and reported that the pod was drifting away at 6 feet per second. By the time it was 2,000 feet away the range-rate had reduced to 4 feet per second and by 3,300 feet it had slowed to 3 feet per second. As it could readily be seen in the dark, Cooper rehearsed manoeuvring to hold it fixed against the background of stars, to stay on a specific line of sight, which was something that would have to be done during a rendezvous, but this proved tricky because the lamp illuminating his optical reticle was so bright that his eye could not adapt to observe the stars beyond.

An artist's depiction of the Radar Evaluation Pod which was ejected by Gemini 5 to test the spacecraft's radar.

Meanwhile, Conrad had been keeping an eye on the fuel cells. Suddenly, he noticed that the *oxygen* pressure had fallen to 260 psi. "How the heck did it get that low?" When told to operate the hydrogen heater manually, he had left the oxygen heater on 'AUTO'. "I think I had better pump that up a little bit." He activated the heater manually, but because doing so required holding the switch in the 'ON' position, he gave it only a short boost. By the time he returned his attention to the radar, the pod was 4,200 feet away and its rate had slowed to 2 feet per second.

As soon as Carnarvon picked them up, Lewis saw the telemetry from the fuel cells and asked them to activate the oxygen heater. Cooper replied that it was in 'AUTO', but did not mention that Conrad had briefly intervened. Just before he lost them, Lewis asked them to operate it manually.

Now in a trouble-shooting frame of mind, Conrad blipped the hydrogen heater and used the amperage meter to verify that it was drawing current, but when he did so for the oxygen heater there was no response – the heater had failed. Although the heater was nothing more sophisticated than a narrow conduit enclosing a wire whose resistance to electrical current generated heat, without it they would not be able to sustain the tank's pressure, and without pressure to deliver the oxygen to the fuel cell, the system would starve and the spacecraft's systems would be denied power.

On seeing that the pressure was now 170 psi and still falling, Cooper reflected that he had never seen a fuel cell run at such a low pressure in the factory tests and, worried that it would totally fail, he decided on a partial power down to relieve the load on the fuel cells.

"We're in the process of powering down the spacecraft," Cooper informed McDivitt a few minutes later, during a brief contact with the Canton Island relay site. He reported the falling oxygen pressure. "You guys think about it for a while."

Hawaii was next in line. As soon as he established contact, Bill Garvin asked about the oxygen. After reporting that it was "160 psi and falling," Conrad told the story – the punch line being, "the conclusion is that we've lost the heater". Hoping that the pressure would stabilise, Garvin told Conrad to keep an eye on it. "What would you like us to do about the REP?" Conrad enquired.

"There's not much we can do about it," observed Garvin. The pod was to test the radar, but the radar had been powered down.

"We were right on the flight plan up until 2 hours and 45 minutes or so," Conrad noted, which was just before they left Carnarvon behind, but now, barely 3 hours into the mission, their plan was obsolete. Doodling idly on the console between their seats, Cooper expressed his mood by drawing a Conestoga wagon balanced on a precipice.

When Guaymas picked them up a few minutes later, Conrad advised the local CapCom, Ed Fendell, that the hydrogen heater, on 'AUTO', was maintaining the hydrogen pressure, but the oxygen pressure was 150 psi and slowly falling. In response, Cooper had powered down some more systems to further reduce the load on the ailing fuel cells. The *good* news was that the pressure sensor data in the telemetry indicated 40 psi higher – apparently the onboard gauge was erring on the low side. Ten minutes later McDivitt asked them to rotate the spacecraft to face the adapter towards the Sun in an effort to increase the 'heat leakage' *into* the oxygen tank, in the hope of increasing the boil off and thereby the pressure feeding the fuel cells. "Power down to the minimum ECS condition," McDivitt called out as they flew out over the Atlantic for the third time.

"It looks like the rate of decrease is decreasing," McDivitt pointed out encouragingly on re-establishing contact via Ascension Island, but Conrad relayed through Tananarive on the island of Madagascar that it was down to 95 psi and still falling. This prompted the urgent request that they adopt "minimum power configuration". If the fuel cells were to starve of oxygen, they would have no option but to revert to the batteries and come home at the first opportunity. By Carnarvon, the gauge was in the high 80s, and by Hawaii it was nearing 65. However, during the fourth orbit the pressure stabilised at about 60 psi. However, they had switched off the UHF voice transmitters to further save power, and were sending telemetry only for brief periods as requested by the ground stations.

As they flew on, the REP traced out a figure of eight, periodically rising and descending and dodging to either side of them, with its range varying between a few thousand feet and a few miles. "We should be rendezvousing with that little rascal," Conrad moaned to Cooper when he saw the pod on the horizon in front of them. "It'd be something if we ran into him, won't it!"

Half an hour later, crossing the Pacific, Cooper noted that the oxygen pressure seemed to have stabilised in the high 50s, and asked Conrad, "Do you suppose they'll have the courage to run us for eight days like this?"

"Boy, I don't know," Conrad admitted.

"I kind of doubt it, Cooper decided.

Concluding that they were probably going to be recalled, they started to pack up their equipment for a '6–4' splashdown northeast of Hawaii in three hours' time, a contingency zone in which the only recovery ship was the oiler *USNS Chipola*.

"I must be unlucky," Conrad speculated.

"Oh, we haven't had really bad luck," Cooper rebuked, "not yet."

As they flew into range of Guaymas, Chris Kraft called: "Gemini 5, Houston Flight."

"Go ahead," Conrad replied.

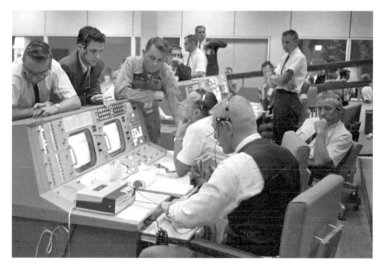

A review of the ailing fuel cell system at the flight director's console: left to right, leaning on console are Dick Glover and John Aaron (both EECOMs) and Elliot See, with Gene Kranz, Chris Kraft and John Hodge seated; leaning on the console in the background are Neil Hutchinson (hand to mouth), Arnie Aldrich (white shirt) and Mel Brooks, with Network Controller Ernie Randall (standing, arms folded on the right).

When Kraft had heard of the power down, he had reasoned that because Gemini 5 had a good battery there was time to hold out for the sixth orbit and see how things developed. He had consulted with See, whose speciality in the Astronaut Office was the fuel cells, and then asked McDonnell to explore their characteristics at the low pressure now experienced. Tests showed that the fuel cells would continue to function at well below minimum specifications, but with correspondingly reduced output. Now up to speed, Kraft laid out his understanding of the situation: "It looks like we have got a situation here that's stabilised, Pete, and we've been discussing the problems associated with the purges. It looks like we can go a fairly long time without any purge." With the oxygen pressure so low, it would be a long time before a build up of impurities would threaten the reaction. "Secondly, it looks like we can purge the hydrogen without any problem. In terms of the oxygen purge, we will probably purge very briefly, not drain the pressure. So I'd like your opinion on going through *another day* under those circumstances." One positive aspect was that as the 190 pounds of oxygen in the tank was used the rate of boil-off would increase and the pressure would rise, but "we're going to have to go a *long time* with you guys sitting up there doing nothing – taking the chance that the fuel cells are going to operate under these conditions for a long period."

"We might as well give it a try," Conrad replied.

And so they flew on in a powered down state, leaving the spacecraft to drift in attitude, with the transmitters silent most of the time. When they were hailed by Hawaii, Garvin told them to continue to purge the hydrogen as planned, but not to purge the oxygen until it rose above 200 psi, a level that now seemed unlikely ever to be recovered. Meanwhile Kraft had decided that with a 13-hour supply on the battery, even if the fuel cells failed they would be able to finish a full day and return to the prime recovery zone in the Atlantic near Bermuda. Garvin relayed this news, "You're 'Go' for 18–1." When Guaymas picked them up Fendell advised them that if they lost the fuel cells they should go to the batteries and draw as little power as possible until it was time to prepare for re-entry.

During the long Pacific crossing, Conrad had been pondering what might have prompted the heater to fail, and he gave Fendell his analysis. While installing the 16-mm movie camera on the first orbit, he had noticed that the circuit breaker for the fuel cell heaters had 'popped' and reasoning that he must have struck it with the camera he had pushed it back in. He now realised that it must have popped when the oxygen heater had blown earlier on. The heaters in the two tanks were on the *same* breaker, so this had disabled the hydrogen heater, and the effect had shown up first in the hydrogen, which was why Carnarvon asked him to keep an eye on the hydrogen and manually operate its heater. In fact, thinking back, he recalled that the oxygen pressure had also been a little on the low side, but not alarmingly so. Ironically, moments later, he had cheerfully noted that they were right on the flight plan! It was a cruel world.

Having made the critical decision to continue, Kraft handed over to Gene Kranz for the second shift, and went straight across the street to the Press conference in order to explain what had happened.

Touch and go

By now, Gemini 5's ground track provided few communication opportunities with the World-Wide Tracking Network and the ships were filling the gaps, but with their radio off there were few calls. Some nine hours into the mission, taking advantage of the lull, Cooper and Conrad had their first meal. Like their predecessors, they had been assigned alternating sleep cycles to ensure that one of them was awake at all times in case a problem developed. Cooper retired first, but got only two hours' of fitful rest. Although it was not practicable to doff their suits, they had removed their helmets and gloves. At five foot six inches in height, Conrad was the shortest of the astronauts, which gave him an advantage in the cramped cabin. With the power off, it was soon so cold that the water vapor in their breath froze on the windows.

Towards the end of Kranz's shift, John Hodge and Kraft decided that because the oxygen pressure had risen to 76 psi as heat leaked into the cryogenic tank, Cooper should check out the spacecraft's systems one by one. To everyone's delight, the fuel cell system carried the load without any sign of distress.

As the 24-hour point passed, Conrad was thinking positively. He came on the radio and announced, " I just told Gordo a few minutes ago, we've just passed a milestone – we only have seven more days to go!" By now, Kraft was confident that the fuel cells would last the duration, and he advanced the goal to '33–1'. However, he remained conservative and when the flight dynamics team proposed a sequence of manoeuvres through which Gemini 5 could be steered in a rehearsal of the early part of an orbital rendezvous he refused, saying that if such a plan were to be put into effect it would have to be later in the mission, and even then only if the power situation had significantly improved. Nevertheless, he did authorise a test of the radar. The REP's battery had long since gone flat, but there was a backup pod at the Cape and this was placed on an unused gantry for the test. On the Stateside pass ending the 17th revolution, the spacecraft's radar and computer were powered up. The radar locked on at a range of 200 nautical miles, as hoped. As the spacecraft flew in towards Florida across the Gulf of Mexico, the closest point of approach was 164 miles. The range and range-rate readings were compared with measurements by a ground radar at the Cape; the spaceborne radar proved to be accurate. As the pressure in the oxygen tank was still increasing, the fuel cells were able to carry the load, the greatest amperage since the onset of the crisis.

"We're in pretty good shape," Conrad pointed out when the pressure reached 86 psi. A few hours later, while Cooper was asleep, Cooper exceeded his 34 hour 20 minute Mercury mission duration. Conrad promised to congratulate him once he awoke. In fact, this was the first time that Cooper had managed to sleep soundly, and got nearly 5 hours of recuperation. Meanwhile, Conrad occupied himself with such tasks as the power situation and the drifting attitude allowed. They had taken on 17 experiments to fill the time. "I hope we can get it all done," Conrad mused as he worked through the list. As it was primarily an endurance flight, their experiments were mostly biomedical in nature, but there were also photographic tasks and visual acuity tests and the cramped cabin was awash with cameras, lenses and rolls of film. When Cooper had reported being able to see fine detail during his Mercury flight, his reports had been received with skepticism, but this time he had a witness. In one experiment they were to attempt to identify patterns on a grid of 2,000-foot squares on open ground 40 miles north of Laredo in Texas, nearby the Mexican border. Because Cooper had said that smoke trails were readily seen from orbit, smoke pots had been positioned to cue them onto the site but the experiment was marred by persistent thin cloud.

After Gemini 4's problem with trash, Cooper and Conrad had planned how they would store frequently used apparatus, and shoved trash such as food packaging down behind their seats. If they had been permitted to open the hatch, they would have jettisoned this garbage mid-way through the flight to clear the clutter. But even if an EVA had been scheduled, this would have been deleted in paring back the flight plan to cope with the power crisis.

As the third day began, Paul Haney, the Public Affairs Officer, reported that "things are looking up" because the oxygen pressure was now almost 100 psi. When McDivitt gave the go-ahead for '47–1', he remarked that Conrad was particularly chatty this morning. Conrad offered to sing him a song. Although Cooper warned that he sang off-key, McDivitt took the bait and Conrad sang him a ditty: "Over the ocean, over the blue, here's Gemini 5 singing to you!"

With the improving power situation, Kraft had authorised the sequence of manoeuvres which, as he explained them to the reporters, were to simulate chasing a 'phantom' target in an orbit ranging between 124 and 184 nautical miles. On a real rendezvous, the Agena target vehicle would have been inserted into a circular orbit at 161 nautical miles but Gemini 5 did not have the propellant to tackle that scenario, and so a more attainable final orbit had been chosen. The significance of this test was that it would be the first time that a spacecraft had manoeuvred *purposefully* in orbit.

For the first manoeuvre, which was over New Mexico a few minutes short of the T+60 hour point, Cooper oriented the vehicle 'blunt end forward' (BEF) and burned the aft-firing thrusters for 28 seconds to slow down by 20 feet per second.

"Right on the money," Conrad reported to Fendell in Guaymas when the burn was over. A radar at White Sands confirmed that the retrograde manoeuvre had lowered the apogee to 168 nautical miles. As they climbed, Cooper yawed the spacecraft through 180 degrees. At the apogee over the Indian Ocean they repeated the burn. A few minutes later they made contact with Carnarvon, whose radar confirmed that the perigee had been lifted to 99 nautical miles. On passing west of Hawaii, Cooper yawed the spacecraft's nose perpendicular for another burn, this one slightly smaller, in order to adjust to the inclination of the spacecraft's orbit with respect to the equator. The result was relayed to Houston by the *Wheeling*, stationed north of Midway Island in the Hawaiian island chain. Once the radars of the Eastern Test Range had confirmed their new orbit, the final prograde burn was performed on the next pass over the Indian Ocean. Whilst the manoeuvre sequence could not be as rewarding as drawing up alongside the REP, because it involved *flying* in space it was the highlight of the mission so far. Jerry Bostick's flight dynamics team in The Trench were delighted. The spacecraft had manoeuvred into an orbit which was not only co-planar with that of the notional target but, by ranging between 108 and 168 nautical miles, was maintaining a fixed differential height about 16 miles below it – a condition known as 'co-ellipticity'. While the spaceborne REP test had been frustrated, this manoeuvre sequence and the test using the transponder at the Cape had made a tremendous step towards the Program's objective of orbital rendezvous. It was good news to Wally Schirra and Tom Stafford, who were in a simulator at McDonnell rehearsing chasing an Agena.

After the rendezvous test, the spacecraft was powered back down again and returned to free drift. At the start of the next orbit they were invited to turn to face the Cape to observe the launch of a Minuteman missile, but they were 1,000 miles to the south and their oblique viewing was through cloud, and they were unable to see anything. The rest of the third day was spent on experiments. When Elliot See asked what it was like in orbit, Conrad replied, "After another day or two, I'll be glad to trade with you." When Kraft came on the line and asked how they felt, they assured him that they were "fine". When Kraft assured them that they were doing a great job in the circumstances, Conrad said that on the revised flight plan they had a busy day of experiments ahead, and they would try and get it all done, to which Kraft said, "Do what you can, that's all we want."

Although the cabin was now a pleasant 70 °F, they were still suffering because the suits were cold. It was not so bad when they were active and the suit was about 50 °F, but when they tried to sleep their immobile bodies did not put out so much heat, and the suit got so cold that they had resorted to disconnecting them from the coolant loop in an effort to stay warm. As Conrad put it, "We've both been sitting here shivering." And it was not just the cold that was keeping them awake. "It's so darned quiet in the cabin, and when one guy is trying to sleep and the other guy does anything, it makes quite a bit of noise." Cooper had managed an average of only five hours of sleep for the first three days. Conrad had fared a little better with 6.5 hours. Nevertheless, because they were now so worn down, sometimes they *both* dozed off during slack moments.

At the start of the fourth day the fuel cell oxygen had increased to 115 psi, and with the trend continuing upward McDivitt gave the go-ahead for '62–1'. It was another day devoted to experiments. In one case, they reported on the test of a rocket sled at Holloman Air Force Base in New Mexico. Driven by 100,000 pounds of thrust, the sled travelled 35,000 feet in 20 seconds and was slowed by a water trough. The objective of the test was to determine if the infrared sensor of one of the onboard experiments could detect it. An hour later, as they approached the coast of California, a Minuteman was launched from Vandenberg Air Force Base just north of Los Angeles. The launch itself was masked by an overcast, but as soon as the rocket emerged from the cloud it became visible. "I see it!" reported Conrad delightedly. Its progress could be tracked by the distinct smoke trail. The missile continued to climb after its engine shut down, peaking far above the Gemini spacecraft at an altitude of 500 nautical miles as it pursued a ballistic arc to its target in the Pacific, some 5,000 miles away. As they completed the pass over the United States, they yawed around to re-run the radar test using the REP at the Cape. Later in the day, they made infrared observations of Kilauea, the most active of Hawaii's volcanoes, which had just stirred back to life.

"Gemini 4, Gemini 4, Houston, over," called McDivitt, momentarily confused when he congratulated Gemini 5 on achieving the half-way point in the mission. Although he relayed the go-ahead for '77–1', a new problem was fast developing involving the fuel cells. The oxygen pressure was now up to 140 psi, but the system was yielding 20 per cent more 'wastes' than expected. The gaseous waste could be vented to space, but the excess water was a problem. The water tank had a bladder to divide the potable water from the acidic waste from the fuel cells. The plan had been for the crew to consume water at a rate matching the output of the fuel cells, but there was no way to vent the excess and the worry was that it would 'drown' the reaction in the fuel cells, causing them to fail.[3] The vehicle was therefore powered down in

[3] This omission in the design would be rectified for later missions.

order to reduce the load on the fuel cells and thereby the rate of production of water, and the astronauts increased their consumption of potable water so as to make room in the tank, with *their* excess water being vented to space via the urine dump system.

As they flew over the prime recovery zone, Conrad tried to find the aircraft carrier *USS Lake Champlain*, but the weather was poor and he was unable to spot it. "I thought an old Navy guy like you could find a carrier!" McDivitt teased. In fact, Conrad had made his first ever deck-landing on this carrier ten years previously. On approaching the California coast 45 minutes later, they spotted another Minuteman launch and this time because the weather had improved they were able to see the missile emerge from its silo.

As they crossed the continent, they claimed the American record for a single flight – set by Gemini 4 – of 98 hour and 32 minutes. McDivitt offered his congratulations. Dr Charles Berry, the Flight Surgeon reported that their biomedical telemetry looked "really excellent", and was providing a valuable continuation of that from their immediate predecessors. When Cooper quipped that they felt a lot better since taking their suits off, Berry jumped up from his console in astonishment before realising that he had been 'had'. After wryly noting that his own pulse rate had soared with the joke, Berry arranged for Kraft, McDivitt and himself to be wired with biomedical sensors for the spacecraft's re-entry.

The cryogenic oxygen tank for the Environmental Control System had to be periodically purged of boil-off. At T+108 hours, it automatically vented and on encountering vacuum the gas instantly froze, sending a blizzard of ice crystals shooting by the windows. "It was really quite a sight," Conrad told Jim Fucci on the *Rose Knot Victor* off the Chilean coast. The display was particularly striking because they were running into orbital sunset.

Pete Conrad inside the spacecraft.

Struggling on

"All set for another day?" Kraft asked at T+116 hours.

"Oh, yeah," Conrad assured. A few moments later he came back, "Gordo and I figure we've been up long enough now to need a 'sim' on re-entry – to get brushed up."

"Do you mean this is the real thing?" Cooper added laconically. "I thought we'd been in the simulator all along!"

On the Stateside pass two hours later, they repeated the radar test with the REP at the Cape, but although the radar locked on this time it would not provide range data, even after Cooper yawed to break and reacquire the signal. In fact, during this activity Cooper realised that the OAMS system was malfunctioning. On contacting Carnarvon Conrad reported that they had "a little problem". One of the yaw thrusters was "not working at all" – it had put out "great globs of liquid" but not delivered any thrust. Actually, the entire system was now "exceedingly sluggish". Suspecting that the solenoid valve had frozen while they were drifting with the OAMS heaters off to save power, they put the heaters back on but it made no difference – on the contrary, several hours later the matching yaw thruster failed as well. With the thrusters consuming propellant without delivering the desired result, Kraft rewrote the flight plan for the rest of the mission to conduct only those experiments which could be done in free drift. The OAMS would be reactivated only intermittently in order to damp out the build up of excessive rates (partly the result of venting the cryogenic tanks). In doing so, Cooper inferred from the increasingly complex cross-coupling of the remaining thrusters that the system was continuing to degrade.

At 119 hours and 6 minutes, they claimed the record that had been held since June 1963 by Valeri Bykovsky. "At last, huh," noted Cooper, on being told that the United States was finally ahead on space endurance. See wished them "Godspeed for the rest of your mission".

Two hours later, on flying over the Atlantic for the 77th time, Conrad reported, "I see a carrier and a destroyer steaming straight into Jacksonville." This time the weather was clear, and the ships stood out clearly. A few moments later, and a few hundred miles further east, he was back, "I think I see either the recovery carrier – or another large ship – making a big wake."

"You're a real homing pigeon for these aircraft carriers, aren't you," observed McDivitt. But as Haney explained in his commentary shortly afterwards, Jacksonville maintained that Conrad had seen a tug towing a barge "which might easily have been interpreted as a carrier and a destroyer to a Navy pilot like Pete Conrad". Nevertheless, in terms of the acuity test, spotting a tug and a barge at sea was an excellent observation.

Now well into the second half of the mission, Cooper and Conrad were becoming tired. As a treat, during long silent Stateside passes, Houston started to play music for them. As both men were 'Dixie Land' fans, when they passed over New Orleans they were provided with a stirring rendition of 'When The Saints Go Marching In'.

"How did you like that?" asked McDivitt afterwards.

"It was great!" Conrad replied enthusiastically.

Ironically, by T+140 hours although the fuel cell oxygen pressure was back up to a very healthy 175 psi, the water situation was growing worse despite the fact that the vehicle had been powered down. Although there was a contingency plan for returning a day early if the power situation declined again, the astronauts were determined that because they had come this far they would not 'Bust'. Nevertheless, it was no fun attempting an eight-day flight in such conditions. They spent much of their final day composing rhymes, which Conrad duly sang to the succession of ground stations.

"We're beginning to feel the effects," Conrad informed See.

"Effects of what?"

"Of being confined for so long – we're getting stiff, and so forth."

"Maybe you ought to open the door and stretch a little bit," See joked.

"I'd sure like to," Conrad observed longingly. The prospect of him making an EVA had expired long before lift-off.

On being given the go-ahead for '122–1' by Dave Scott they wryly observed that there was a storm heading for their prime recovery zone.

When McDivitt took over again, he said that he had visited their wives earlier and they had watched Gemini 5 pass overhead. Jane Conrad had been moved to write a poem, which McDivitt proceeded to read:

Twinkle, twinkle, Gemini 5, how I want you back alive.
Up above the world so high, I saw you today as you went by.
Twinkle, twinkle, Gemini 5, tomorrow you take your great big dive.
Zinging towards the ocean blue and I send my love to you.

"Tell her I think that's great," Conrad replied.

Having accomplished most of the experiments that could be made in light of the power situation and free-drift, the flight plan for the final day was sparse. Cooper and Conrad became bored. Although well paired, they had long since swapped their best stories, and so they whiled away the hours in silence. Cooper became particularly testy. When Dr Berry asked whether they were getting enough exercise he replied sardonically, "I hold Pete's hand once in a while. I used a cleaning towel. And for a couple of days we chewed gum." And then he added even more sarcastically, "We thought we'd start taking long walks." Conrad later said that he wished that he had brought a book to read.

At T+177 hours Kranz decided to bring them home one revolution early so as to avoid the storm (now upgraded to Hurricane Betsy) that was threatening the nominal zone – the new site, designated '121–1', was several hundred miles further west, halfway between the Cape and Bermuda. Several hours later, as Gemini 5 entered its 114th revolution, Arda Roy on the *Coastal Sentry Quebec* stationed off Japan pointed out that this was their final pass over his ship and he wished them a nice landing. "Thanks for all your help," Conrad replied, "you did a real fine job."

As a distraction for the astronauts as the spacecraft began its 117th revolution, Houston relayed a signal down the Eastern Test Range to Antigua, which transmitted it to Gemini 5.

"This is SeaLab II transmitting from 200 feet down off La Jolla," called Scott Carpenter. "How do you read, Gordo?" Carpenter, a Mercury veteran, had just entered the lab on the floor of the Pacific on a mission to study sea-life. His voice was a high falsetto because the atmosphere was pumped up with helium. The radio was 'noisy', but Carpenter and Cooper were able to swap pleasantries and Carpenter signed off by wishing them a good re-entry.

On revolution 118 Conrad noted that the needle on the PQI meter had slipped below the zero marker. Cooper confirmed that the OAMS tanks were dry by firing a thruster with no response. The frozen valves had evidently been leaking propellant. Shortly thereafter, Berry reminded Cooper that it was his wedding anniversary and he passed up congratulations from Trudy Cooper. In preparation for the de-orbit sequence, Cooper powered up the Re-entry Control System. Kraft pointed out that if they had to abandon the retrofire for '121–1' they should hold off for '123–4' in the Pacific.

"Everything is just peachy-keen," Cooper assured Fendell at the Guaymas station. "It's nice to have a control system again." He had spent much of his Mercury flight drifting in a powered down state, and now, to his dismay, had done so again on Gemini 5. For the final phase of the mission, the batteries had been brought on line so that the spacecraft could be powered up.

"The old fuel cells have done very well," Cooper told McDivitt later on the pass across the United States.

"They sure have," McDivitt agreed.

This remarkable test flight had revealed that the fuel cell system was more robust than its designers had believed, and could run at a pressure far below that which had been taken to be its minimum – which was nice to know.

The de-orbit and re-entry information was uploaded into the spacecraft's computer by the Retrofire Officer, Tom Carter. At McDivitt's recommendation, the 'fail safe' burn had been elided.

Short landing

Gemini 5 drew its record-breaking eight-day mission to an end on 29 August. In the dark, high above Hawaii, Cooper held the spacecraft stable and Conrad threw the switch to jettison the equipment section of the adapter module. Garvin counted down for retrofire to provide a check of the spacecraft's timer. The first three retros fired in

After the swimmers had attached the flotation collar, the astronauts egressed and waited for the recovery helicopter to lower a harness to retrieve them one at a time.

rapid sequence, but there was a lengthy pause before the fourth one ignited. On emerging into daylight over the Mississippi, Cooper verified his instruments against the horizon. At the entry interface at 400,000 feet he initiated a series of roll manoeuvres, first banking one way and then the other in an effort to refine the length of the trajectory and to correct any cross-range errors. When the computer indicated that they were too high and were going to overshoot, he increased the scale of his banking to 90 degrees to reduce lift and thereby shorten the trajectory, in the process increasing the dynamic loading almost to that of a ballistic re-entry, which was stressful after such a long period of weightlessness. Whilst the computer continued to show that they were coming in 'long', when they splashed down at 07:56 Cape time they were 70 nautical miles short! An investigation found a flaw in the calculations for the de-orbit burn – the onboard computer had been mislead, with the result that Cooper's efforts to correct its false indication served only to progressively draw them short of the recovery zone.

The *USS Lake Champlain*, the prime recovery ship, was 80 nautical miles further east, but the destroyer *USS Du Pont*, having been stationed 72 miles uprange of the carrier, was very close by. Nevertheless, because they were so far off target it was 45 minutes before a helicopter from the carrier arrived and dropped swimmers to attach the flotation collar. The sea state was calm and the astronauts initially elected to remain with their spacecraft, but on being told that the carrier was several hours away they changed their minds. On hearing that Cooper and Conrad were safely on the recovery helicopter, Kraft broke out his cigar box for Charles Mathews and Robert Gilruth, who had come in for the final moments of this historic mission. When the astronauts reached the carrier an hour and a half after splashing down, some of the medics expected that they would be so weak as to require to be carried, but they walked along the red carpet – although because their legs were rather wobbly they did link arms for mutual support. Looking tired but pleased with themselves, the two men playfully tugged on their beards. Once below deck, they were subjected to several hours of medical tests, with X-rays being taken of their chests and heel bones, their blood chemistry sampled to determine the volumes of plasma and red cells, EKG readings taken, eyes tested, and neuro-motor tests performed.

Displaying their 8-day whiskers, Conrad and Cooper make their way across the deck of the *USS Lake Champlain*.

"There is absolutely nothing wrong with them," was the immediate medical report. "It's just as though they had taken a short plane ride." Nevertheless, ongoing medical monitoring found both Cooper and Conrad to be dehydrated and somewhat weakened, as well as being decidedly tired because the chilly temperatures had undermined their attempts to sleep. One factor in Cooper's state of exhaustion was dubbed "commander's syndrome", reflecting his greater sense of responsibility for attaining his mission's objectives. The loss of blood plasma and bone calcium was more pronounced than on previous missions, but they soon recovered in both deficiencies. Whilst Gemini 5 had shown that a crew could survive weightlessness for long enough to fly a circumlunar mission, and the extra time required to mount an expedition to the lunar surface seemed unlikely to be impractical, the trends in the body's adaptation to weightlessness did raise the prospect that a very long flight (such as a tour of duty in a space station) would face problems.

Afterwards

Despite the power and attitude control constraints, only one of the 17 experiments had been cancelled;[4] the others were at least partially successful. It turned out that the valves on the thrusters had clogged. One contribution that the Program was to make

---

[4] The Near Object Photography experiment was to have been done after rendezvousing with the REP, and so could not be pursued.

to Apollo, was a reliable reaction control thruster system. It was only by identifying how the early versions failed during long duration missions that the design could be refined. Without this testing on Gemini, some of the early Apollo missions may well have been recalled by thruster failures.

Having finally wrested the endurance record from the Soviet Union with a 190 hour and 55 minute mission, in mid-September President Johnson dispatched Cooper and Conrad on an international tour which included a visit to the International Astronautical Federation in Athens, where they met Pavel Belyayev and Alexei Leonov, who had flown Voskhod 2 in March. Deke Slayton disapproved of these post-flight distractions because it interfered with feeding the astronauts back into the rotation. While Slayton had tended to preserve backup crews, rotating them to prime slots three missions downstream, he assigned Armstrong to command Gemini 8 and See to command Gemini 9. Cooper moved to Apollo development. Conrad was made backup commander of Gemini 8, which put him in line for command of Gemini 11 with the likelihood of returning to space within a year – *operational flying*.

Chapter 5

Rendezvous In Space
Gemini 6/7

Rendezvous strategies

When NASA realised that it would have to become proficient in orbital rendezvous, there was considerable debate regarding the degree to which the spacecraft ought to fly itself. The engineers wanted a fully 'closed loop' system in which the computer would accept the radar data, compute a manoeuvre, orient the spacecraft, and make the burn. The astronauts were wary of an automated system. What if the radar failed? What if the computer failed? Surely the crew were not passengers! Their presence onboard would provide redundancy against a hardware failure. It was decided to employ an 'open loop' strategy in which the pilot would decide whether to allow the computer to proceed with a burn. When the Space Task Group established the 'Capsule Review Board' in January 1961 to determine whether it would be possible to modernise the Mercury capsule to undertake orbital manoeuvres, this concluded that rendezvous was "too hazardous for a one-man operation". A new spacecraft would be necessary, with a crew of two, with the Command Pilot flying the vehicle and the Pilot, who would be seated alongside, operating the radar and the computer.

When Gemini was conceived, orbital rendezvous was a daunting challenge. Few people knew *anything* about it, and those that did knew that it would be difficult.

It takes energy to lift a payload against the Earth's gravity. It takes more energy to put a satellite in a higher orbit than it does into a lower one. By Johannes Kepler's laws of orbital dynamics, vehicles in lower orbits travel more rapidly than those higher up. Manoeuvring in orbit is a matter of *energy management*. For a spacecraft to raise its orbit, it must accelerate; to lower its orbit, it must decelerate. Either way, it will expend energy, and hence propellant. In the 1920s Walter Hohmann made a theoretical study of orbital dynamics and discovered that the most efficient means of rendezvousing involves flying a minimum-energy 'transfer orbit' in which the chase vehicle travels one-half-leg of an elliptical orbit that intersects both orbits tangentially. When Gemini planning began it was presumed that this method would be pursued, but by early 1964 four options were under consideration:

* direct insertion
* tangential
* co-elliptic

* first apogee

In the *direct insertion* technique the active vehicle would be put into an orbit as nearly identical to that of the target as practical, and in close proximity to it, and the active vehicle would then use its own engine to close in. Although this might appear straightforward, if the launch vehicle were to deliver off-nominal performance then the active vehicle might not be able to close the gap in either the time or the propellant available. The Soviet Union showed in 1962 that it could insert one spacecraft close to another, but as its Vostok spacecraft was incapable of manoeuvring in space the difficulties of interception were not explored. In any case, direct insertion was merely a special case of rendezvous in which the chase started on the ground – it did not address the general issue of one vehicle in space rendezvousing with another, and so NASA did not pursue it.

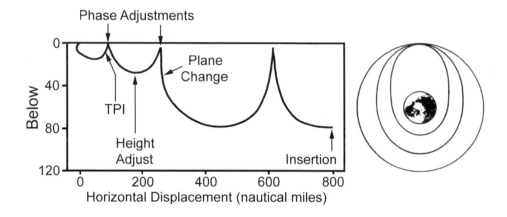

The tangential rendezvous scheme. Note that the overview (right) is not to scale.

In the case of *tangential* rendezvous, the target would be placed into a circular orbit and the active spacecraft launched as the target completed its first orbit, and be inserted into an elliptical orbit having its apogee tangential to that of the target. After a series of manoeuvres calculated by the ground on the basis of radar tracking, the interception would be conducted using onboard guidance while rising to apogee on (say) the third orbit. This method had two major drawbacks. Firstly, as the active vehicle would be inserted directly into the tangential ellipse the rendezvous would occur at apogee, which for a launch from Florida would be on the opposite side of the world and in the southern hemisphere, and because it was a Program requirement that launches be made in daylight this would mean that the interception would take place in darkness – which was deemed to be unacceptable. Secondly, the establishment of the initial conditions for the interception was very sensitive to what mathematicians call 'dispersions' (that is, position and velocity errors) in the ground-controlled manoeuvres, and hence unforgiving because by the time these errors became apparent they would be difficult to correct.

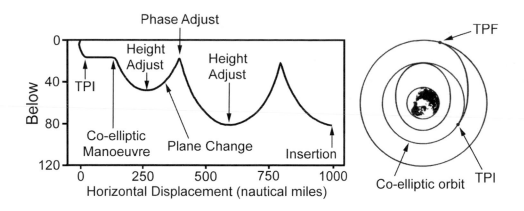

The co-elliptic rendezvous scheme using a 130-degree terminal phase.
Note that the overview (right) is not to scale.

For *co-elliptic* rendezvous, the ground-controlled manoeuvres would be similar to those for the tangential approach, but rather than setting up the interception they would establish the active vehicle in a co-elliptic orbit that maintained a specific difference in height just below the target's orbit. If the active vehicle was placed behind the target, it would slowly catch it up. This offered the flexibility of selecting the lighting conditions for the interception. At a time determined by onboard guidance, the active vehicle would perform the 'terminal phase' of the rendezvous in which it would climb to intercept the target.

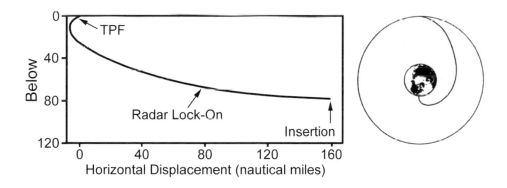

The first-apogee rendezvous scheme.

In *first-apogee* rendezvous, the launcher would place the active vehicle into an elliptical Hohmann transfer orbit which could either be circularised at its apogee to draw alongside the target or provide the jumping off point during the approach to apogee for a short co-elliptic transfer using onboard guidance. This scheme posed two drawbacks. Firstly, the early part of the insertion orbit was to be reserved to enable the crew to check out the spacecraft, and it was considered impractical to initiate the onboard guidance during this time. Secondly, the rendezvous was not only extremely sensitive to the dispersions at orbital insertion but also unforgiving of out-of-plane errors. In fact, the in-plane and out-of-plane consequences of a lift-off that was a mere five seconds off the ideal moment would impose a penalty of about 300 feet per second, and as this represented a significant fraction of the likely manoeuvring capacity of the Gemini spacecraft this would render the rendezvous impractical.

In 1964, when the issue of which rendezvous strategy to use for Gemini was resolved, it was considered unwise to expect to launch within a five-second window, and this ruled out first-orbit rendezvous. In any case, this would have been rejected as the *primary* method of rendezvous, because if it failed it would be necessary to transition to another method for a second attempt at rendezvous.

The preliminary manoeuvres for the tangential and co-elliptic methods were similar, but the lighting imposed on the interception by tangential rendezvous was unfavourable. As the closing rate in the terminal phase of a co-elliptic rendezvous would be *slower* than when the active vehicle was rising to apogee in the vicinity of its target, it was more tolerant of errors.

Dean Grimm of the Flight Crew Support Division and Buzz Aldrin, the Astronaut Office's representative on the rendezvous team,[1] successfully argued for co-elliptic rendezvous. But this was only the start, because if the *theory* was to be put into practice, *procedures* would have to be developed.

Optimisation studies enabled the key parameters to be refined. One factor was when to initiate the terminal phase. This involved trading the differential in height between the orbits against the catch-up rate and the range at the initiation point. The larger the differential, the more rapid the catch up, and thus the less time for the crew to prepare after entering the co-elliptic orbit. It was decided that the greatest range at which the crew could be expected to locate the target visually in sunlight was 50 nautical miles. Some time would then be needed to observe it optically. With the target in a near-circular orbit at 161 nautical miles, and the transfer initiated at a slant range of 30 to 35 miles, a difference in height of 15 miles would yield a transfer subtending a central angle of roughly 130 degrees. As the active spacecraft caught up, the target would appear to 'rise' in elevation and the time to initiate the transfer would be when it was 27 degrees above the local horizon. Because the terminal phase was to be initiated with the Gemini spacecraft aimed at its target, this condition would be measured in terms of the pitch angle.

The primary constraint was that the terminal phase occur in darkness in order to enable the Command Pilot to maintain the target both centred in his optical reticle and fixed against the background of stars. (This, of course, required the target vehicle to be equipped with an acquisition light.) Meanwhile, the Pilot would monitor the

---

[1] When Aldrin was recruited as an astronaut, he already had a doctorate from MIT on rendezvous theory.

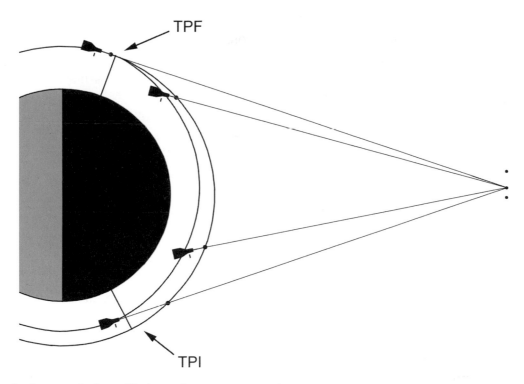

In the canonical co-elliptic rendezvous scheme, the terminal phase transfer is initiated shortly after the target enters the Earth's shadow, the spacecraft's nose is aimed at the target, and the inertial frame of reference for making the midcourse corrections is established by maintaining the target fixed against the stars. The target is in a near-circular orbit at 161 nautical miles and the active spacecraft is 15 miles lower. As the active spacecraft catches up, the target rises in elevation. The transfer is initiated when it is 27° above the local horizon, and will subtend a central angle of about 130 degrees. Whereas at TPI the Gemini spacecraft is trailing at a slant range of about 33 miles, it continues to pitch up and travel along the line of sight and passes through vertical to end up inverted and stationary a few hundred feet in front of the target.

radar's range and range-rate, and operate the computer to provide a series of course corrections. An additional constraint was that the braking phase (during the final few miles of the approach) be in daylight for safety. These constraints placed the Terminal Phase Initiation (TPI) manoeuvre shortly after orbital sunset, and the interception point just after sunrise. As for when the active spacecraft was to manoeuvre into the co-elliptic orbit, it was expected that a crew would require about half an hour (and probably much longer on the first attempt) to prepare for the terminal phase. The canonical plan by which the active spacecraft would be steered into an orbit co-elliptic with and a constant 15 nautical miles below the orbit of the target, involved *four* manoeuvres.

Whilst the second stage of the Titan launch vehicle would attempt to insert the Gemini spacecraft into an elliptical orbit with its apogee at the required differential below the target, there was the option of a 'height adjustment' at the first perigee to establish the 15 nautical mile differential.

Half an orbit later, at the newly established apogee, there would be an opportunity for a 'phase adjustment' in which the perigee would be raised to establish the desired phasing (i.e. timing) for the co-elliptic manoeuvre.

So long as the out-of-plane error inherited from the launcher was within the capacity of the spacecraft to correct (for the Gemini spacecraft, the limit was half a degree) it would be possible to make a 'plane change' manoeuvre at one of the two 'nodes' at which the orbits intersected.

Finally, with the two orbits both co-planar and with their major axes aligned, at the third apogee the co-elliptic manoeuvre would raise the perigee to establish the constant differential in height below the target's orbit. If the target's orbit was perfectly circular, then this would be a circularisation, but the target's nominal 161 nautical mile orbit would likely vary from this by a few miles and the objective of the ground-controlled manoeuvre sequence was to *match* this ellipticity.

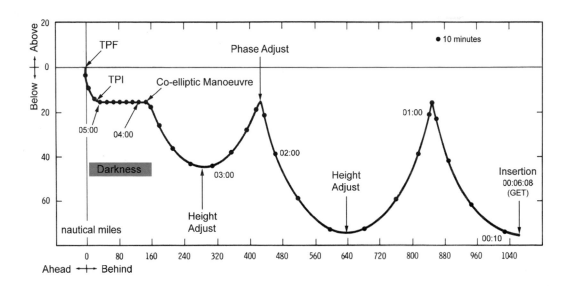

The details of Gemini 6's 'm=4' rendezvous with Gemini 7.

To allow time for the World-Wide Tracking Network to measure each orbit and then to compute the next manoeuvre, and also to allow the spacecraft's crew a significant period in the catch-up phase to prepare for TPI, it was decided to make the interception on the fourth orbit. It could not be left much later than this, because as the Earth rotated the ground track would migrate away from the stations of the tracking network. Using the terminology of the flight dynamics specialists, a rendezvous on the fourth orbit was referred to as 'm=4'. Once this had been perfected, it would be possible to tighten up the timing for faster procedures.

The Gemini-Agena target vehicle

In August 1961, when James Chamberlin drew up the specifications for 'Mercury Mark II', for which rendezvous would be the primary operational requirement, the options for the target vehicle were to modify either the Air Force's planned SAINT 'satellite interceptor' or the Agena upper stage. The SAINT vehicle was several years from flight trials.[2] The Agena, however, was not only already in service, it was by far the most reliable of the upper stage available. Furthermore, the Agena was compatible with the Atlas launch vehicle, with which NASA was already familiar. Upon being established in January 1962, the Gemini Program Office formally selected the Agena. The Air Force's Space Systems Division managed the contract with the Lockheed Missile and Space Company to build the Agena. NASA requested twelve vehicles,[3] and by June it had been agreed that Lockheed would supply a modified version as the Gemini-Agena Target Vehicle (GATV).

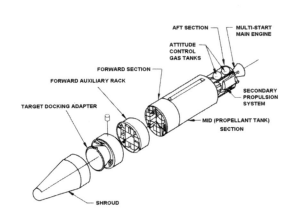

The major structural components of the Gemini-Agena Target Vehicle.

In addition to using the Agena to put satellites into orbit and to dispatch probes into deep space, the Air Force had integrated some of its payloads directly into the vehicle, so NASA decided to mount the Target Docking Adapter (TDA) on the Agena's nose. In addition to a conical docking mechanism, this adapter would incorporate a pair of acquisition lights and a radar transponder. NASA also requested a number of significant modifications to the Agena. Firstly, it wanted the engine to be capable of igniting up to half a dozen times to enable it to perform extensive manoeuvres in orbit, in some cases while docked to a Gemini spacecraft.[4] NASA also added a secondary propulsion system, upgraded the attitude control system, and augmented the stored program controller with a communications system so that an astronaut in a Gemini spacecraft could send commands to the vehicle. In its GATV configuration, the Agena's Secondary Propulsion System (SPS) comprised a pair of engines set in 'saddle bag' fashion, one on each side of the 16,000 pounds-thrust Primary Propulsion System (PPS). In fact, each of these engines had a pair of thrust chambers, one delivering 16 pounds of thrust and the other 200 pounds of thrust, and the corresponding pairs were fired in unison. Major manoeuvres would be made by the PPS, and minor ones by the larger units of the SPS, with the smaller units making such 'tweaks' as may be required. The Bell Aerosystems Company of Buffalo in New York set about modifying the PPS for the multiple restart capability by replacing the solid-charge starter which generated the gas to drive the turbine with one that generated the

---

[2] The Radio Corporation of America was awarded the contract to develop SAINT in 1960. It would combine a propulsive stage and a payload designed to inspect satellites. The project was cancelled in December 1962.

[3] The initial request for a dozen Agenas reflected the expectation that one would be required for each Gemini mission, but funding constraints limited the number of Atlas boosters that could be made available and so this order was halved.

[4] The early models of the Agena could fire their engine only once, to burn into orbit. The new 'D' model that had only just come into service when NASA became interested in the vehicle could fire its engine twice to be able to make a follow-on burn to send its payload to its operating station. Unfortunately, the *ad hoc* means by which this second ignition was achieved was not capable of being extended to further burns.

gas by vaporising liquid drawn from a small tank. Unfortunately, the Program got off to a shaky start due to an early cash flow crisis and, rather surprisingly, the 'pacing item' became the Agena's PPS upgrade. By the end of 1963, Charles Mathews ordered the Aerospace Corporation to make a systems analysis of the project, and to mediate technical issues. Nevertheless, a few months later, the GATV contract was at risk of cancellation. In April 1964 the Air Force recommended flying a test vehicle to assess its operational status, but NASA refused because this would have used up an expensive Atlas (these were also in short supply at that time). Instead, NASA decided to subject the first vehicle to roll off the production line – designated GATV-5001 – to rigorous ground testing. This propulsion unit emerged from preliminary trials in late 1964, was fitted with its TDA in January 1965, and then both the primary and secondary propulsion systems were comprehensively tested.

Rendezvous crew

In April 1964, a week after the test launch of the Gemini Launch Vehicle, Deke Slayton assigned Wally Schirra and Tom Stafford to back up Gus Grissom and John Young for the 'shakedown' mission because this would give Schirra and Stafford the best possible start to prepare for the first rendezvous. Although the crewing for Gemini 6 was not announced, in February 1965, during the preparations for Gemini 3, Schirra leaked to reporters the fact that he was to command the first rendezvous and docking mission. In making the official announcement in April, Paul Haney, the Public Affairs Officer in Houston, jokingly teased the reporters that he was about to reveal one of the agency's "best kept secrets".

The terminal phase of the rendezvous was the tricky part, because the crew would have only their radar, the computer and an optical reticle. Although both vehicles were pursuing differently curved trajectories, by maintaining the spacecraft facing the target and the target fixed against the stars, it would be possible to fly a *straight line* to the target. Or to put this more technically, the transfer would be made using an *inertial* frame of reference, with the active vehicle maintaining a constant inertial attitude – a procedure that greatly simplified the midcourse manoeuvres. On closing within a few miles, and while continuing to maintain the target fixed, the Command Pilot would brake on a profile designed to halt alongside the target. In addition to rehearsing the ideal case, it was necessary to explore how to recognise and recover from various flaws.

McDonnell constructed a simulator in its plant in St Louis in which Schirra and Stafford developed procedures for the terminal phase of the co-elliptic rendezvous by trial and error. Each simulator 'run' concluded with a thorough assessment to determine what worked and what did not. While a computer would be used to compute the rendezvous, the crew would take responsibility for braking, station-keeping and docking because there was no provision for automatic docking – the *pilot* was therefore a key component in the system. To assist in training, a Gemini mock-up was mounted on a special rig and small thrusters fired to induce manoeuvres to line up the spacecraft with the Agena and ease its nose into the conical TDA collar at 1 foot per second in order to trigger the three capture latches; the docking would be completed by 'rigidising' the collar with the nose of the Gemini spacecraft locked within it.

The Langley Research Center slung a Gemini mock-up on a
mount that provided six degrees of freedom to enable astronauts to
rehearse docking with an Agena (as this multiple exposure depicts).

Spacecraft 6 was the first off the McDonnell production line to be specifically adapted for EVA, but Schirra wanted to focus the flight plan on testing rendezvous and docking and he ruled out a spacewalk by Stafford as a distraction. Furthermore, since his spacecraft was powered by batteries, and so had limited endurance, he wanted to fly a straightforward up-and-down mission. The mission was slated for 48-hours, with the proviso that if it achieved all of its objectives in the first 24 hours it would return at that point. Even though Gemini 5 had rehearsed a sequence of manoeuvres leading to co-ellipticity, Gemini 4's unexpected difficulty in returning to its Titan stage had prompted skepticism of achieving a rendezvous on the first attempt.

Setback

In May 1965, GATV-5002 was delivered to the Cape and mated with the Atlas that was to launch it as the target for Gemini 6. As this required coordinating parallel countdowns, it promised to be the trickiest mission yet attempted by the Cape's launch teams. In an effort to gain some experience, a 'dual countdown demonstration' had been held involving two unrelated launches. On Monday, 25 October, the counts for Gemini 6 converged. The Atlas was on Pad 14 and the Titan II was on Pad 19, a mile further north along Missile Row. The objective was to launch the Atlas at 10:00 local time and send the Gemini spacecraft after it about 100 minutes later, the exact timing depending on the parameters of the target's orbit.

The Atlas-Agena for Gemini 6 on Pad 14. Notice the gantry has not yet rolled back.

By about T–100 minutes the 30-minute operation to roll away the large Gantry Service Tower was concluded. Over the next 40 minutes the Agena's tanks were filled, firstly with hydrazine fuel, then with nitric acid oxidiser, and pressurised. At T–50 minutes, the Chief Test Conductor at Pad 14 was assigned control of the master clock. Until the Atlas's launch, any holds that might be requested by Cape Kennedy, Mission Control in Houston, or by the World-Wide Tracking Network would be declared by Pad 14. Ten minutes later, the liquid oxygen was pumped into the Atlas – a 20-minute task. The refined kerosene fuel had been loaded the previous day. As the cryogenic fluid would boil off, a valve was left open at the top of the tank to vent the gas, and liquid was pumped in to keep the tank full. At T–24 minutes the Eastern Test Range's Range Safety Officer verified his radio command link to the Atlas, which he would use to trigger the destruct system if the Atlas veered off course. This was followed by a series of final telemetry checks of both the Agena and the Atlas. At T–8 minutes, the Agena switched to its batteries. At 3 minutes and 30 seconds the Atlas's telemetry system switched onto internal power. At 2 minutes and 30 seconds there was a final check of the systems in the launch vehicle. Twenty seconds later, the gaseous oxygen vent on the Atlas was shut and the tank pressurised. At 1 minute and 40 seconds the Atlas went onto internal power, and water was pumped onto the pad at the rate of 30,000 gallons per minute. At 1 minute, the range safety destruct system in the Atlas was armed.

A 20 second hold began at T–18 seconds, during which the Test Conductor monitored a panel of lights that turned from the yellow to green as the final key events occurred. When the last green light came on, he enabled the automatic sequencer to pick up. At the 4-second mark, the ignition sequence began. The two vernier engines on the side of the Atlas ignited first. When they attained their operating pressure the side-mounted boosters and the central sustainer engine at the base of the vehicle ignited and built up a total of 395,000 pounds of thrust, with the vehicle lifting off 3 to 4 seconds after the initiation of the ignition sequence, precisely on time. The Atlas recognised the fact that it was airborne by a motion switch that actuated when the booster was 2 inches off the pad. It thundered into the sky and arced out over the

The Atlas-Agena lifts off on 25 October 1965.

Atlantic. The booster engines shut down after 2 minutes and 11 seconds, and were jettisoned. The sustainer cut off two and a half minutes later.[5] Once the vernier engines had refined the trajectory, the launch vehicle released its payload to continue on a ballistic arc.

After some 50 seconds of coasting, the Agena fired the two 16-pound-thrust 'cold gas' engines of its Secondary Propulsion System for ullage to settle the propellants in their tanks for the firing of the Primary Propulsion System, and 6 minutes and 20 seconds after lift-off the sequencer issued the signal to ignite the PPS for the 3-minute burn that would insert the vehicle into a 161-nautical-mile circular orbit. Jerome Hammack, NASA's Agena Officer in the blockhouse on Pad 14, reported that the telemetry stream stopped after the briefest sign of ignition. His counterpart in Houston, Mel Brooks, saw the same thing. At that same time, the Flight Dynamics Officer noted that the Eastern Test Range's radar was tracking several objects. In the blockhouse, Colonel L.E. Allen, the Air Force officer in charge of the launch, suggested that the vehicle had exploded.

Wally Schirra and Tom Stafford were already strapped into their spacecraft and eager to give chase. It was decided to continue the clock just in case the Agena had indeed made it to orbit. However, hopes fell when Canary Island failed to hear the radio beacon. Carnarvon in Australia was the key, because it had a radar capable of tracking the vehicle by 'painting' it. When Carnarvon reported "No joy!" NASA accepted that the Agena's engine had exploded and scrubbed the Gemini 6 countdown.

Having egressed their spacecraft, Wally Schirra leads Tom Stafford back down the ramp on Pad 19.

---

[5] For the Titan, SECO means 'second-stage engine cut-off', whereas for the Atlas, it means 'sustainer engine cut-off', but in each case it signifies the time at which the launch vehicle shuts down.

Rapid replanning

Frank Borman and Jim Lovell had taken a break from training for Gemini 7 and flown to the Cape to watch the day's events. They had selected a vantage point to the north, on the Merritt Island Launch Area that the Press was already referring to as 'Moonport USA'. The Atlas's launch had been spectacular. On hearing that the Agena had failed, they made their way to the Pad 19 blockhouse. On their way, they happened across Walter Burke and John Yardley, two of McDonnell's managers, who were discussing the possibility of mounting a dual mission in which Gemini 6 would rendezvous with Gemini 7. Yardley knew that the Martin Company had proposed "a rapid-fire launch demonstration" of two Gemini missions. Pad 20 had been built for the Titan II missile, and could be reconfigured for Gemini–Titan launch vehicles. In September 1964, George Mueller had asked Bill Schneider, the Gemini Mission Director, whether it would be necessary to upgrade Pad 20 in order to sustain the planned bimonthly flight rate, and Schneider had said that a single pad should be sufficient. Nevertheless, the existence of the second pad had set Schneider thinking and in February he had proposed "salvo firing" the final two missions of the Program and having one astronaut from each spacecraft swap places with this counterpart in orbit as a grand finale.[6] Eager to preclude a hiatus while the Agena problem was rectified, Yardley called over Raymond Hill, McDonnell's chief at the Cape, and asked him whether he recalled an idea raised by Joseph Verlander, his Martin counterpart, to demonstrate a "rapid turnaround" by launching Gemini 6 as soon as possible after Gemini 5. Verlander had proposed stacking the launch vehicle for Gemini 6 on Pad 19, checking it out (a time-consuming process) and then using a Skycrane helicopter to transfer it bodily for storage on Pad 20. After Gemini 5 had been launched and the pad refurbished, the stored vehicle would be moved back and mated with its spacecraft. However, this scheme had been rejected as an unnecessary complication. Gemini 7, which was scheduled for December, would not require an Agena, but it was evident that any delay to Gemini 6 would have a serious knock-on effect on the schedule for the remainder of the Program. Listening to this discussion, Borman promptly endorsed the proposal for mounting a joint mission. It would not only enable his mission to launch on schedule, it would make it more interesting. It would be feasible only because Gemini 7 was to be a 14-day endurance mission – the longest of the Program – but the deciding factor would be the time required to refurbish the pad and stack and prepare Gemini 6. Could it be done within 10 days? Burke and Yardley sought out George Mueller and Charles Mathews, who were also at the Cape. After the slow pace of the Mercury Program, the prospect of achieving and sustaining a 60-day turnaround in order to fly five Gemini missions in 1966 had seemed daunting, they said, and so the proposal to launch two missions 10 days apart was ridiculous. Undeterred, Burke and Yardley joined the post-mortem in the nearby Manned Spacecraft Operations Building. The consensus there was to switch the immediate effort to Gemini 7, and ground Gemini 6 until the Agena was fixed. Each

---

[6] Schneider's study was a month before Alexei Leonov surprised NASA by making the first spacewalk, at a time when the official line was that the first Gemini EVA, which was due in the summer, would involve no more than swinging the hatch open.

spacecraft was configured for a specific mission, so it was initially hoped that it would be possible simply to replace Spacecraft 6 with Spacecraft 7 on GLV 6, but then it was noted that the 14-day version of the spacecraft would be too heavy, for the Titan on the pad. At the conclusion of the meeting, Mueller telephoned James Webb in Washington and proposed slipping Gemini 6 to February or early March 1966, by which time the Agena problem should have been rectified. He also noted that the crew's training schedule precluded advancing Gemini 7. Webb accepted this recommendation.

The next morning, after flying to Houston to continue to lobby for their plan, Burke and Yardley put it to Robert Gilruth, who was skeptical.

"Tell me what's wrong with it?" Burke challenged.

Gilruth called in his deputy, George Low, who was intrigued by the idea but doubted the World-Wide Tracking Network would be able to handle two manned spacecraft, as opposed to a Gemini and an Agena. Chris Kraft was summoned for an opinion. As they were waiting, they were joined by Charles Mathews, who had already judged the plan to be unworkable.

"You're out of your minds," said Kraft, "it can't be done." When he was unable to put his finger on the fatal flaw, he said that he would consult his operations staff. Typically, the flight directors were primed to identify and overcome flaws in plans. The WWTN issue was resolved when John Hodge suggested that they handle the passive vehicle using the network that had been developed for the Mercury missions, in which the individual ground stations monitored the spacecraft's status during its overflight and teletyped a summary to Mercury Control; there had been no real-time distribution of telemetry. Hodge argued that by the time that Gemini 6 was ready for launch, Gemini 7 would have settled into its routine, and that in addition to relaying the telemetry from Gemini 6 to enable the flight controllers in Houston to monitor its progress during the rendezvous, the remote stations could send a teletyped report on Gemini 7's status. Kraft ordered his Mission Planning and Analysis Division to develop a flight plan for a mission such as that being proposed, knowing that any 'fatal flaw' would be highlighted by the rigour of the planning process. Deke Slayton sounded out the two crews. Borman reaffirmed his support. Schirra was also enthusiastic. Having touched all the bases, Kraft informed Gilruth that the plan was viable so long as the pad could be refurbished and Gemini 6 prepared in 10 days. After discussing the matter again with Mathews, Gilruth called Mueller, who was now in Washington, and told him that he was in favour of the dual mission. The next day Mueller asked Robert Seamans to recommend the proposal to Webb. Seamans duly put it to Webb and his deputy, Hugh Dryden that afternoon.[7] Webb sought a firm recommendation from Mueller, who called Gilruth, who in turn polled Kraft, Slayton and Merritt Preston, who was Houston's chief at the Cape. In the absence of any 'second thoughts', Gilruth discussed it with Mathews. Mueller relayed their support to Webb as his formal recommendation. After reviewing the pros and cons, Webb decided in favour, and arranged for the President's office to make the public

7 Unfortunately, Hugh Dryden died of cancer on 2 December, and so did not live to see this historic mission.

announcement, and the next day Bill Moyers, President Johnson's Press Secretary, announced that in December Gemini 6 would attempt to rendezvous with Gemini 7 in orbit. Impressed by the plan's audacity, the Press promptly dubbed the mission '*Spirit of 76*'.

Remarkably, within 60 hours of the Agena blowing up, NASA had developed a far more ambitious mission. This improvisation marked the agency's rapidly growing confidence in its spacecraft and the supporting infrastructure. The decision was an organisational triumph. At almost every step along the way, people had overcome their initial skepticism and set out to make the plan work. Furthermore, at no stage as the proposal had advanced up through the levels of management had its originators, and the pilots who would fly the mission, been left out. This institutional recovery from the loss of the Agena was NASA's management at its best.

In Joseph Verlander's proposal, a helicopter would have moved the fully stacked Titan from one pad to another in order to satisfy Aerojet's requirement that the stages remain in a vertical orientation, but it was now decided to relax this constraint and so once Spacecraft 6 had been demated and stored in a satellite hangar on Merritt Island, the Titan was destacked and its stages trucked for storage on Pad 20, a few hundred yards north up Missile Row. On 29 October GLV-7 was erected in its place, and the lengthy checkout procedure started in a mood of mounting expectation.

The plan for Gemini 7

Slayton had inserted Frank Borman into the rotation early so that he would be available for the 14-day mission, because he thought that Borman was sufficiently tenacious to see it through. His initial position was backup to the Gemini 3 crew, but the assignments were reshuffled and Borman switched to backing up Gemini 4, which at that time was to have tackled the 7-day mission, and teamed with Jim Lovell. On 1 July 1965, after Gemini 4's return, they were rotated to Gemini 7, with Ed White switching to backup Borman and Michael Collins being brought in to backup Lovell.

Realising that spending two weeks in pressure suits which could not be removed would be taxing, in early June 1965 Gordon Cooper and Elliot See performed an altitude chamber test wearing Air Force flight suits, helmets and oxygen masks, and reported enthusiastically that this was far more comfortable than wearing a G3C suit. The counter argument was that the crew would need full pressure suits to be able to eject at high altitude during the ascent. James Correale of the Crew Systems Division proposed a lightweight pressure suit similar to the G3C but not rated for vacuum. Instead of the metal neck-ring and attachable fibreglass shell helmet, this 'soft' suit would have an integrated hood which zipped into position, and the wearer would use a conventional pilot's helmet for communications. This 16-pound suit would be able to be removed in space. However, McDonnell expressed concern about cabin temperature and pressure failures. Furthermore, in the event of a fire the procedure was to vent the cabin's air. If the crew had doffed their suits, what were they to do? On the other hand, tests showed that the Environmental Control System operated *more efficiently* when the crew were unsuited. Nevertheless, it was decided to develop such a suit for the 14-day endurance flight for crew comfort, but to Cooper's chagrin it was not made available to his mission. It was initially planned that if the spacecraft was healthy at the end of the first full day, both crewmen would strip to their longjohns, but Mueller and Seamans demanded that one man be suited at all times in case of an emergency. Starting with Lovell, therefore, they were to take turns every 24 hours. When the rendezvous was added to the flight plan it was decided that as a safety precaution they would both suit up for this exercise.

Having backed up Gemini 4 and closely followed the progress of Gemini 5, Borman and Lovell had soaked up all the lessons learned by their predecessors. On Cooper's mission for example, storing trash had become a serious issue, and this would only be worse on a longer flight. Since Borman and Lovell would be wearing the 'soft' suits, they would not have the option of opening the hatch to jettison their accumulated garbage. It was decided to discard the first week's food packaging by stuffing it behind Borman's seat, and the second week's trash behind Lovell's seat – it was a case of 'out of sight, out of mind'. Another lesson from Gemini 5 was that no one knew how long any given task would require in weightlessness (it was not just the performance of the task, but the preparation and packing up afterwards) so there was little point in planning a detailed timeline. An 'outline' flight plan was developed instead. This listed the tasks to be addressed, specifying their external constraints. The non-time-critical tasks were to interleaved with the time-critical events as conditions allowed. As it was to be a 300-hour mission, there was a fair chance that they would 'tick all the boxes'. One advantage of this approach was that the addition of the rendezvous did not seriously disrupt this flight plan. Furthermore, such flexibility was considered likely to keep this crew in a better psychological state than their predecessors who, on finding themselves 'behind', had grown frustrated trying to 'catch up' on an unrealistic schedule. And, of course, there was the issue of sleep. The alternating sleep cycles had been a disaster. But Gemini 5 had established that the spacecraft – even if it was ailing – could safely be left untended while the crew slept. With confidence in the systems, Borman and Lovell decided to start out on the same cycle, which would be synchronised with Houston time, and see how things developed.

At times during their eight-day mission Cooper and Conrad had grown bored. In part, of course, this was due to the chilly conditions resulting from the power crisis. One of Conrad's suggestions was that Borman and Lovell each take a book to read if the mission turned sour. Borman took 'Roughing It' by Mark Twain, and Lovell chose 'Drums Along The Mohawk' by Walter Edmonds, which was about the American pioneers. Of course, having accepted a full programme of experiments, if the mission went to plan there would be no time to read.

Of the 20 experiments, eight were biomedical in nature reflecting the fact that, to a much greater extent than previously, Borman and Lovell were test subjects for an evaluation of the human body's reaction to the space environment. The biomedical experiments were labelled M-1 to M-9 (M-2 was not carried). For M-1, which addressed cardiovascular conditioning, Lovell (as had Conrad) was to wear rubber cuffs that would inflate periodically to compress his thighs in order to stimulate the autonomic nervous system, to assess whether this would reduce his reaction to weightlessness and hasten his recovery upon return to Earth. Borman (and Cooper) served as 'controls' for the test. M-3 utilised an inflight exerciser. This was a bungee cord with a foot strap and a hand loop, and when extended to 12 inches it applied a force of 65 pounds. It was designed to exercise muscles in the arms and legs. Since it was an experiment, not just an exerciser, its effect was to be monitored by noting heart rate prior to and following a specified number of calibrated pulls, and then the time taken for the heart to recover its previous state recorded. M-4 was the inflight phonocardiogram that measured the sound of the heart to study possible fatigue of the heart muscles; the heart is a muscle in its own right, remember. M-5 was a bioassay of body fluids. It measured the body's output of fluids. In the first few days in weightlessness, blood pooled in the upper body because there was no force to pull it down into the legs, and receptors in the thoracic area interpreted this pooling as being an *excess* of body fluid and induced increased urination to compensate. The objective of this experiment was to monitor this process over a prolonged time. During their flight Conrad and Cooper suffered significant demineralisation in some of their bones. M-6 investigated this by taking X-rays of the heel and bones in the little finger prior to and after the mission. M-7 addressed calcium imbalance. It involved monitoring calcium intake and output prior to, during, and after the mission in order to determine precisely how the body released calcium while exposed to weightlessness and recovered it on return to Earth. Since calcium could be lost in urine, feces and perspiration, measuring its output was a significant imposition on the crew because it required capturing *all* bodily wastes, including the sweat soaked up by the longjohns that were worn under the suit. In fact, to create baseline data to assess the effects of the flight, Borman and Lovell and their backups were on controlled and monitored diets for several weeks leading up to launch, and even during this run-up period they wore urine and feces collectors, wore a single pair of longjohns, and bathed in distilled water that was retained for study. M-8 required Borman having EEG electrodes glued to his scalp to document his brain activity for a study of "the sleep-wakefullness cycle, degree of alertness and readiness" in the space environment.

The Gemini 7 patch expressed the marathon nature of the mission.

When Ed White made his spacewalk on Gemini 4, some of the medics had expected that he would suffer severe vertigo, but he had reported feeling "a million dollars". M-9 was to study the sense of balance provided by the otoliths of the vestibular system of the inner ear which, being gravity sensors, are unable to function in weightlessness. Some of the astronaut corps resented the discomfort imposed by such 'fishing trips' for the effects of space flight, and did not envy Borman and Lovell their fortnight as lab rats for the medical community. Their other experiments included studies of the radiation environment in space, the Earth's magnetic field, and photography.

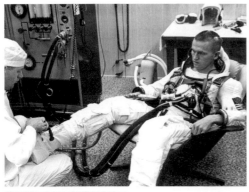

Preparing for Gemini 7. Frank Borman in the suit-up trailer on Pad 16. Note the sensors on his scalp (top). Al Shepard has a final word with Borman, who has donned the hard pilot's helmet worn under the soft-suit (centre). Wearing their sealed soft-suits, Borman and Lovell walk up the ramp on Pad 19 (bottom).

In effect, with these long duration missions NASA was learning how to live and to work within the very confined environment of a spacecraft – there was nothing like experience for learning lessons; it was pure 'human factors' research. The 'marathon' analogy was reflected in the mission patch, which depicted a flaming torch.

Gemini 7 sets off

At 07:00 on Saturday, 4 December, Borman and Lovell were awakened by Deke Slayton. During breakfast with Slayton, Al Shepard, Wally Schirra and Tom Stafford, Gus Grissom and John Young, Pete Conrad and Dick Gordon, and Neil Armstrong and Dave Scott, Lovell said that he was "looking forward to a good long flight". Meanwhile, Ed White and Michael Collins, the backup crew, prepared the spacecraft. Borman and Lovell were in good spirits when they arrived at Pad 16, where they donned the 'soft' G5C suits that Lovell had helped to design. As they ingressed, they found an embroidered patch with the words '*Home Sweet Home*' that had been placed in the spacecraft by Collins. After the hatches were sealed, the White Room was vacated and the erector lowered. The scattered showers were clearing up. After Al Bean, the 'Stoney' CapCom, worked it through a remarkably smooth countdown, Gemini 7 lifted off on schedule at 14:30.

"We're on our way, Frank!" Lovell observed.

"Right!" Borman agreed.

Schirra and Stafford watched from the roof of the blockhouse on Pad 37, which was a Saturn IB facility further north up Missile Row, and White and Collins watched from a hut on the beach.

The first stage's thrust was about 2 per cent on the low side, but as they approached

'Point 8', Elliot See, the CapCom in Houston, assured them that they were "right down the slot".

"Bingo!" exclaimed a delighted Borman at SECO, some 5 minutes and 40 seconds after lifting off. The perigee of 87 nautical miles was nominal. Whilst the apogee of 177 nautical miles was six miles lower than planned due to the first stage's shortfall, this was not a serious problem.

"That's the best 'sim' we've had," congratulated See.

After the requisite 30 seconds, Lovell threw the switch to detonate the pyrotechnics that jettisoned the spent stage and Borman immediately fired the thrusters to move clear. Since it had been decided that they should rerun the station-keeping exercise that had frustrated Jim McDivitt, Borman fired the spacecraft's thrusters for only a few seconds to preclude straying far enough to permit orbital dynamics effects to draw the two vehicles apart – the plan was to remain within 100 feet of it – and then promptly cancelled his motion. Meanwhile, Lovell affixed a 16-mm movie camera in his window to document the experiment. "Let's see if we can see that son-of-a-gun," Borman said to Lovell as he began the turn. "There she is, Jim! There she is!" Then to Houston he reported, "We have the booster in sight. It looks pretty."

"Roger, booster in sight," See acknowledged.

"It's venting drastically – really venting." Excess propellant was being released through a valve near the engine. There was some venting from the engine nozzle, too. The venting was acting like a thruster, causing the stage to tumble end-over-end at a rate of several times per minute. Few people had seen another vehicle adrift in space, so the sight was something of a novelty.

"*Slowly* venting?" See queried.

"It's *really* venting!" Borman emphasised.

"Watch it, Frank!" Lovell warned, when it looked like they would stray too close.

"It's okay," Borman assured.

"How's the station-keeping going?" See asked.

"It's going very well," Borman replied. In fact, he was having to fire the thrusters rather more than he had expected in order to stay with the stage, since the venting was inducing a translation component as well as a rotation, and it was moving away from them. "Fantastic sight, isn't it!" he said to Lovell on the intercom.

"Great!" Lovell agreed.

"Are you getting pictures?"

"The Sun's in the way, Frank."

Unfortunately, as they had launched in the afternoon and were heading east towards the sunset terminator, on turning around to inspect the spent stage they found that it was backlit by the Sun. Although this illumination made the venting appear spectacular, the vehicle was silhouetted, so Borman manoeuvred around to the north for a better view.

"Now I have it," Lovell pointed out, as the stage appeared in the corner of his window about 100 feet away.

Facing south, the Sun still lit up Lovell's window, but Borman's, in darkness, faced the looming terminator.

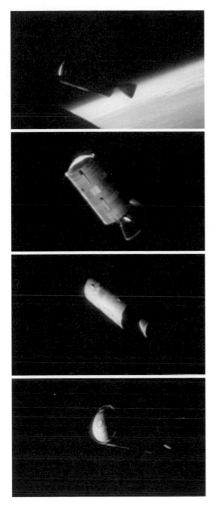

Immediately after attaining orbit, Gemini 7 turned to station-keep with the upper stage of its Titan launch vehicle, which was filmed by Lovell.

With 30 seconds to loss of signal through the Eastern Test Range, Lovell reported that they had finished the post-insertion checklist.

"How is everything over in those fuel cells?" Borman asked.

"Looking good," Lovell assured. The batteries had been isolated, and the fuel cells were carrying the load.

When Canary Island acquired the spacecraft and relayed its telemetry to Houston, Flight Director Chris Kraft told Jim Fucci: "Your data is coming in tagged 'Agena'. That's wrong, it should be tagged 'Gemini'. We'd like for you to tag it right." The remote site was not to treat Gemini 7 as an Agena until later in the mission, when it was being pursued by Gemini 6.

A sharp drop in the oxygen pressure feeding the fuel cells prompted Fucci to ask Lovell to activate the heater. Even with the heater on, the pressure continued to drop. In Houston, See was joined by Cooper – between them they had the most experience with the fuel cells, See having participated in their development and Cooper having observed them operating in space.

Meanwhile, Borman passed the control stick to Lovell to give him a feel for flying the spacecraft, and Lovell closed in to 50 feet of the slowly tumbling 28-foot-long stage. A few moments later, they flew into the Earth's shadow and everything went black.

"I can't see a darned thing," Borman complained. "Can you?"

"No."

Although the Titan stage had been fitted with flashing acquisition lights for this mission, it was still difficult to judge how far away it was. Gemini 7 had a docking light – in effect, a spotlight – but this was not much use unless they turned to maintain its beam on the target. Borman reclaimed the stick and thrusted to move clear.

"Be sure you're far enough away," Lovell prompted.

The flight plan called for a burn to separate from the stage 25 minutes after launch, but because they were already down to the 88 per cent propellant cut-off for the station-keeping test Borman decided to make the burn early. He turned 'blunt end forward' (BEF) to make the manoeuvre, which he shortened "to save some gas".

A minute after they made the separation burn, See called through the voice-remoting site at Kano in Nigeria. "We're standing by for you to burn."

"We made it early," Borman pointed out.

Having studied what had happened to McDivitt when he had attempted to return to his booster, and having spent many hours in the simulator, Borman showed that station-keeping was feasible so long as the range was not allowed to open too far. This served to encourage the Gemini 6 crew, as it meant that once they managed to close to within about 100 feet of their target they would have no trouble.

"Is the oxygen pressure coming up at all?" See asked.

"It's still going down – it's 100 psi now," Lovell said. As a precaution, the fuel cells had been modified to enable oxygen to be fed in from the Environmental Control System's tank. See asked Lovell to open the cross-feed valve, let it rise to 250 psi and then close the valve.

On reaching Carnarvon, Keith Kundel gave them the go-ahead for '17–1'. They reported to Ed Fendell in Hawaii that after the separation burn their PQI was 82 per cent. As a result of the burn Gemini 7 was in an orbit in which, for the next few hours, the range to the stage would vary between a few hundred feet and 10 miles. Over the Atlantic for the second time, heading once more into sunset, Lovell described the stage as "a brilliant body" a little ahead. Flying on, Lovell monitored the fuel cell pressure. When it dipped below 200 psi, he turned on the heater, which this time restored the pressure without recourse to the cross-feed valve. They were in great shape.

Remoting through Tananarive on the island of Madagascar on the third revolution, See asked the two 'rookies' how they liked it so far.

"It's great," Borman replied.

"Outstanding," Lovell agreed.

By now, the spacecraft had been oriented 'small end forward' (SEF) so that a 77-second burn by the 100-pound-thrust aft-firing thrusters could raise the perigee to 120 nautical miles in order to extend its orbital life.

"We *hit* something!" Lovell reported after the burn. "Something came 'fluttering' by the right window. It looked like a strap or a piece of tape. Borman had interrupted the burn for a few seconds when this had occurred, but had picked up and achieved the required change in velocity. When asked later if the strap had come from the front or the rear of the spacecraft, they ventured that it had come from the rear. The consensus was that it was a strip of tape which had protected the pyrotechnic charge on the 'separation plane' where the adapter had been mated to the upper stage of the Titan, and that it was 'blown forward' by the pressure of the efflux from the aft thrusters. Although harmless, the impact had caused a moment of concern.

Meanwhile, with the PQI following the perigee-raising burn reduced to 67 per cent, See asked them to minimise manoeuvring until they could resume the consumption profile for the mission. In Houston, having got the mission off to a good start, Kraft handed over the Flight Director's chair to Gene Kranz. When Gene Cernan, the second shift's CapCom, asked how they were doing, Lovell told him that they were "in the process of settling down for a long winter's flight".

When Charles Lewis on the *Coastal Sentry Quebec* stationed south of Japan asked how they were planning to take their sleep period, he received the reply that they were "giving it consideration". Unlike their predecessors, Borman and Lovell were to sleep at the same time, and more or less in sync with Houston time. Their first sleep period coincided with the 'gap' in the World-Wide Tracking Network which was covered by the ships. Ten hour blocks of the flight plan had been set aside for crew rest. As much as feasible, the monitoring stations were to leave the crew in peace, calling only if some system was in need of urgent attention. After several hours dozing, the astronauts finally fell asleep. At the end of his 8-hour shift, Kranz handed over to John Hodge. The Flight Dynamics Officers, however, had decided to pursue a different cycle. Each would work for 16 hours through the active shifts supervised by Kraft and Kranz and then spend Hodge's slow shift 'on call' on a camp bed in the nearby bunk room. In this way, the three FIDOs would cycle through 24-hour periods in Mission Control and have 48-hours off.

Spacecraft 6 is hoisted back on top of its restacked launch vehicle.

Meanwhile, at the Cape, Pad 19 was a hive of activity. When a rocket lifted off, the pad technicians were generally granted several days of rest prior to starting the clean-up routine, but just 8 minutes after Gemini 7 departed the Martin Company sent in inspectors to assess the damage, which was minimal. Engineers, welders, electricians and plumbers soon went in to refurbish the facilities. At 21:00 the first stage of GLV-6 arrived. Over the next two hours it was placed into the erector, and the erector raised. By 02:00 the stage was on the pad. By 09:30 the second stage had been mated with it. Spacecraft 6 had arrived from its storage at 07:30, and this was hoisted up onto the rocket at noon. "That's what I like to hear!" Kraft said on being informed that the stacking process had been achieved several hours ahead of schedule. When Guenter Wendt, McDonnell's Pad Leader, was told about the need for a fast turnaround he had been dismissive: "Oh, man, you're crazy!" It usually took *weeks* to stack and check a Titan booster, but as the two certified stages had been carefully stored, this time only the interconnections had to be verified. As a result, no sooner had the stack been mated than its checkout began.

The first week

When biomedical telemetry showed that Borman and Lovell were active, Kundel called and found that they were having breakfast after four to six hours of what Borman described as "light sleep". They had slept in their soft suits, but with gloves off and the hood pushed back to serve as a soft headrest for their hard hats. "It's a little warmer than we thought it'd be, based on Pete and Gordo's experience," Borman pointed out. In contrast to the crew of Gemini 5, who had chilled in a powered down spacecraft, Borman and Lovell were running their suits on full-cold to keep cool.

It was a day of experiments with the focus on how the body adapted to weightlessness, and the astronauts soon settled into a routine of periodically reporting their blood pressure, heart rate, temperature, how many ounces of water they had consumed, which meal packs they had opened and which contents they had either eaten or discarded, when they filled a urine bag, and when they felt moved to use what was politely referred to on the radio as a 'blue bag'.

As Gemini 7 completed its first 24-hours in space, with all of the spacecraft's systems operating smoothly, See gave the go-ahead for '31–1'. A little later, in discussing the object that had struck the window during the burn, Ed White, standing in as CapCom, said "We'll send up 'Six in a few days and take a look at it."

"There's an idea!" Lovell agreed.

After a second day of experiments, Bill Garvin on the *Rose Knot Victor* stationed in the South Atlantic signed off for the night, "Now you both go to sleep after our LOS."

"Roger, will do," Lovell promised.

At the Cape, the checkout of the Titan was running two shifts ahead of schedule, and it was beginning to look as if it would, after all, be possible to pull the rabbit out the magic hat and send up Gemini 6 for the rendezvous.

It took several calls to awaken Gemini 7 ahead of schedule to perform a fuel cell purge. The third day was to be a milestone in the Program. For the first time, an astronaut was to take off his pressure suit. By now the cabin temperature was 72 °F, the suit inlet was 58 °F and they were wearing their soft suits with the zippers down in order to keep cool. In this state, the soft suit could be sealed in a minute or so, but once removed it would take at least half an hour to don again. But the medics wanted to collect biotelemetry with one astronaut suited and the other not. After consulting the chief Flight Surgeon, Dr Charles Berry, Kraft approved the experiment.

"How do you feel about taking your suit off?" See asked, but received no reply because the spacecraft had passed beyond the Eastern Test Range. On contacting Canary, Borman reported that Lovell would be "glad to" take off his suit, but he would have to find "a slack period" for the operation, which was expected to take about 10 minutes. On doing so two hours later, Lovell found that in the cramped cabin it took much longer than expected – nearly an hour. He folded the suit and stuffed it down beside his seat, out of the way. An 'orbital flight suit' had been provided, but finding himself to be "very comfortable" in his longjohns he decided not to unpack the suit. "I feel kind of naked without my suit on," he mused.

To enable Gemini 7 to serve as a target for the rendezvous, it had been provided with a transponder like the one on Gemini 5's REP. When the spacecraft flew over the Cape, this was interrogated by a ground radar to verify that it was working properly. When Houston asked for the computer to be kept on after the test, Borman was blunt: "Would you remind Mr Kraft that computers take electrical power, and electrical power uses cryogenics, and we're not going to be able to borrow any [cryogenics]."

"Mr Kraft knows all that good stuff," said Kraft pointedly. "Standby, we want to check your computer. Last look, you have enough cryogenics to stay up there for about 20 days."

"Okay, Chris," Lovell replied on behalf of his commander.

Frank Borman in space.

As a further preparation, music was played on the HF uplink to test the receiver which was to be used when Gemini 6 communicated directly with Gemini 7 without blocking the UHF air-to-ground link. It was not long before the Press dubbed Mission Control the Kraft Music Hall.

The astronauts monitored the launch of a missile by the *USS Benjamin Franklin* off the Cape. It was a Polaris A-3, and they detected it as it rose from the ocean and ignited its motor. "We've got her, and she's beautiful," Lovell reported excitedly. "I hope they know where they're shooting that thing." He had a movie camera in his window, and Borman was slewing the spacecraft around to track the missile's progress. "It's easy to track, we're right on it," Borman reported as he kept it centred in his optical reticle, which was practice for the forthcoming rendezvous. They followed the missile throughout its 2,500-nautical-mile flight down the Eastern Test Range. With the excitement over, they retired. Although he was "a little warm", Borman retained his pressure suit. Lovell luxuriated in his longjohns, and was grateful that he could scratch the itchiness around the biomedical sensors. Although the first night's sleep had been "pretty poor", they were now getting solid rest. Their synchronised routine was proving to be much less stressful than the alternating one, and it helped that the WWTN maintained radio silence.

On Pad 19, meanwhile, progress was such that it began to look as if it would be feasible to launch *a day ahead* of schedule, on day 8. Given this uncertainty, Gemini 7's flight plan had to be flexible. The primary task was to circularise Gemini 7's orbit at 161 nautical miles before Gemini 6 was launched. A requirement was that the WWTN have sufficient time to determine the rate of decay of the 'circular' orbit, because any ellipticity would complicate the set up for the rendezvous.

On awakening, Lovell said that he had kept waking up because he was "slightly cool" in his longjohns, but could not be bothered to unpack his orbital flight suit because of what he called the "housekeeping problem". See read up the newspaper headlines, one of which was '*Lovell orbits in underwear*'. Borman promised to snap some pictures of his pilot "for the album". Meanwhile, Borman gave the medics the bad news that he had torn off some of the sensors on his scalp. "I'm sorry about the

EEG experiment. The harness caught when I was trying to put something away, and ripped them off." In fact, only one night's sleep data had been secured because one of the sensors had worked loose during the second sleep period.

Seventy hours into the flight, a 12.4 foot per second prograde burn tweaked the orbit to provide the option of optimising for either the 8th or 9th day. The PQI was down to 59 per cent, but this was sufficient to achieve the circularisation without jeopardising manoeuvring for the experiments.

By the time that the astronauts retired, the preparation of the Titan was 20 hours ahead, and the Spacecraft 6 checks were nearly finished. "The launch vehicle folks are standing by for the spacecraft to catch up," the Public Affairs Officer observed. At this pace, it would be possible to advance the simulations to certify the spacecraft by 24-hours, and aim for launch at 10:10 on Sunday, 12 December.

When Kraft asked how they were doing after four days in space, Lovell replied, "We're pretty good, Chris. It's amazing, the spacecraft seems to get bigger and bigger – either we're losing weight or we're getting used to it; I don't know which."

As Borman and Lovell worked through their list of experiments, Kraft had the WWTN report their monitoring of the spacecraft using the Mercury teletype message system for six hours. Meanwhile, Houston linked up with the Cape for simulations involving Spacecraft 6. In the first run, Grissom and Young sat in the capsule in their work clothes and simulated a Mode II abort 100 seconds into the ascent. Meanwhile, Schirra and Stafford were suiting up for a simulation in which they performed a countdown from the T–45 minute mark. The aim of such simulations was to 'exercise' the systems to verify the interconnections between the launch vehicle's stages, and between the Titan and the spacecraft.

"Boy, they're really working on Pad 19 down there!" Lovell mused appreciatively on being told the news. "We just passed over the Cape, and with our telescope we could see them working."

"It's very feverish activity," See agreed.

Next time around, Gemini 7 made a brief radio contact with Schirra and Stafford as they worked through an abbreviated simulation of the orbital manoeuvres that they were to make in chasing Gemini 7.

It was at this point that Gemini 5 suffered its first thruster fault, but Gemini 7 was doing better. The OAMS heaters were keeping the system warm – the temperature in the adapter immediately behind the thrusters was typically 50 °F, whereas with the heaters switched off on Gemini 5 this temperature had fallen to 20 °F and the solenoid-valves had frozen. Hopes were high that properly heated thrusters would function throughout the Gemini 7 mission.

The Cape was still aiming for a Sunday launch, but it had been decided that if they were obliged to wait for day 9, the circularisation would be split into two parts, with the first burn raising the perigee and the second lowering the apogee. "If we don't circularise tomorrow," came the news, "that is, if we go for a Monday launch, we'll circularise on Sunday." In fact, there was some concern over the rate at which propellant was being consumed. Some of the experiments that required either turning the spacecraft in a specific direction or slewing it to track a target were drawing more propellant than expected, with the result that the PQI was down to 54 per cent, in which case it might be necessary to curtail some of the experiments.

One hundred hours into the mission, the *Coastal Sentry Quebec* relayed a request from the medics in Houston that Lovell don his orbital flight suit for the next sleep period "to get another data point" on crew comfort. As Lovell was replying, Borman was heard to sneeze. Dr Berry promptly demanded a full report! Borman admitted that he had a dry nose and his throat was a little irritated. Lovell felt fine. Although Lovell put on the orbital flight suit, he stripped it off again after 15 minutes because it was too hot, and slept in his longjohns.

While Borman and Lovell slept, a cycle of end-to-end propulsion system tests were run through the night at Pad 19, and cryogenics loaded into the spacecraft's tanks. The decision for which day to launch was to be made at noon on the coming day, Thursday, 9 December.

Having secured biomedical data from Lovell for three days unsuited, Dr Berry proposed that Lovell don his pressure suit so that Borman could doff his. This would not only permit Borman some comfort, it would also provide an opportunity for Lovell to re-assess the suit. However, Borman declined. "If we have to keep one suit on, I'd prefer to have me keep my suit on and leave Jim the way he is." Borman was smaller than Lovell, and did not find the cabin quite so cramped. If Lovell stretched out with his feet in the footwell, his head rubbed against the hatch, so he had to maintain a seated posture. But the medics had a rationale for wanting Borman out of his suit. When Berry pointed out that Borman was drinking rather more water than Lovell, Lovell volunteered, "I'm not sweating as much as he is." The cabin was 78 °F and even with his suit ventilation on full-cold and his zippers fully down Borman was sweating. "I'm perspiring *a little*," he allowed. Berry persisted, because the results had indicated that the unsuited crewman had a lower blood pressure and pulse rate. Borman was insistent. "We'd like to stay just the way we are until Tom and Wally come up, and then Jim will get back into his suit." The unsuiting and suiting up procedure was time-consuming and there was a risk of upsetting switches in the process, and he saw little point in taking off his suit now because they would require to suit up for the rendezvous. "I'd prefer not to switch and then get back in. Let's go the way we are if we have to stay with one suit off." Houston yielded.

A few hours later they were given the go-ahead for '90–1' and alerted to prepare for the circularisation.

"We'll see if we can fit it into the schedule," Borman joked.

"That's very good of you," See replied.

Asked whether they thought that they could go the full 14 days, Borman described their spacecraft as "a Cadillac", by which he meant that it was running smoothly. He agreed that they would both doff their suits after the rendezvous "until, say, one day before re-entry." And on the general issue of suits, he recommended, "There's no question in our minds now that the only way to fly these things is without pressure suits." Apollo would have to be a shirt-sleeve environment.

A little later, Borman complained, "One of my urine sample bags just came apart in my hand."

"Before, or after?" enquired See.

"After!"

Dr Berry promptly asked whether Borman had managed to recover some of the sample.

"I caught most of it in my face and hands."

After mopping up, they had their lunch and then prepared for the two-part manoeuvre to circularise their orbit. The first burn was made over Australia. It was prograde, with the aft thrusters firing for 1 minute and 18 seconds to add 61.2 feet per second in order to raise the perigee to 162 nautical miles. Borman reported their success via Canton Island. The second burn was to be 180 degrees later, over the Eastern Test Range. As they crossed the United States, See pointed out that they had exceeded the longest Soviet mission of 119 hour and 6 minute. "We're more interested in Pete and Gordo's record," Lovell replied. For the second burn, Borman oriented the spacecraft BEF and fired the aft thrusters for 15 seconds to drop the apogee to circularise the orbit. With the PQI down to 33 per cent, See told them to "go back to being stingy with your fuel again".

"After that, we'll have to," Borman reflected. But as this was the last major manoeuvre they were in good shape.

"It looks like lift-off time for 'Six will be about 09:45 Cape time," See confirmed, which was a reference to Sunday.

While the astronauts slept, Hodge's flight controllers enjoyed a nutcake which had been baked by his wife, and was reliably reported by the Public Affairs Officer to be "delicious". The following day, an impromptu Suit Configuration Meeting decided that Lovell must suit up in order to permit Borman, who now had developed a stuffy nose in addition to his other symptoms, a period of comfort. When Garvin on the *Rose Knot Victor* relayed the decision, Borman remained reluctant.

"You can advise them that Chris Kraft wants him to get back in the suit," Kraft said to Garvin, who relayed the message.

"I'm not comfortable, but I'm alright," Borman insisted. "It's a big job for Jim to get in and out of the suit." But he promised that he would strip down to his longjohns after the rendezvous. On the next pass over the United States, Kraft called Borman directly. Borman reiterated that he would rather stay as he was, but when Kraft observed pointedly, "Frank, I think you understand what's going on down here", he finally relented: "Aye, aye, sir. We'll do it. If you want us to, that's it." But he put it off a little. "We'll change tonight." Borman began the night in his longjohns, but upon awakening in a chilly state he donned *both* orbital flight suits to warm up. Lovell, now suited, was okay. On taking down the covers from the windows in the morning, they found that the cabin had cooled down by 20 degrees because they were tumbling rapidly. On Gemini 3, evaporation from the water boiler had imparted a yaw motion. The system had been modified, but now the fault had reappeared. It was as if the system was simply venting rather than boiling its water. With the PQI down to 26 per cent there was insufficient propellant to spare to keep restabilising the vehicle for a specific experiment. It was decided to null the rates only periodically to keep them manageable and otherwise to drift until the rendezvous.

"It's starting down hill from here," said Ed White, Borman's backup, when he called up later in the day to note that they were half way through the mission. "You're really doing a great job."

"The best decision was when Mr Kraft ordered me out of that suit!" Borman admitted.

"You just made *his* whole day."

"He made my whole day *and night*," Borman mused. "Lovell is without comment!" His pilot was not enjoying being suited.

In fact, despite their strenuous efforts to maintain a 'tidy ship', it was showing signs of having been lived in for a week – and it was not simply the accumulating trash but also the inevitable aroma.

On reclaiming the CapCom's console, See told them that Schirra and Stafford had run a final simulation of the rendezvous in Spacecraft 6 overnight "and everything went real well". A little later he was able to confirm, "Everything is 'Go' for GT-6 tomorrow." On hearing that Borman planned to suit up again in good time, See told him not to do so until *after* the launch.

"Tell Wally and Tom we'll be looking for them."

Tom Stafford in the suit-up trailer (left). Wally Schirra strides up the ramp on Pad 19.

Pad abort

While the Titan was being loaded with propellants, both crews slept. The World-Wide Tracking Network was configured to treat Gemini 7 as if it was an Agena. The countdown started at 05:30 on Sunday, 12 December, at the T–240 minutes mark. While Grissom and Young prepared the spacecraft, Schirra and Stafford were awakened at 05:20, breakfasted with Gordon Cooper, and left for Pad 16 to suit up.

"We've had an excellent countdown, with no known problems," reported Jack King, the Public Affairs Officer at the Cape.[8]

The clock was held at T–3 minutes for 25 minutes to wait for the final tracking, so as to load the launch azimuth as the target passed over. The slight drag of the tenuous ionosphere had drawn Gemini 7 down to a 161.5 nautical mile circular orbit, which was perfect for the rendezvous.

Schirra observed in amusement that there were a number of flying ants on his window, and the Spacecraft Test Conductor, Don Kroner,[9] pointed out that these bugs were in for a surprise and a short ride.

"Gemini 6 is fuelled up and ready to try," Arda Roy at the Corpus Christi site in Texas relayed to Gemini 7 as it flew over heading towards Florida with seven minutes of the hold remaining. "Houston says the weather at the Cape is 'Go' for launch."

"They're cleared for take off," Elliot See informed Gemini 7 as the clock began to tick down the final three minutes of the exceptionally smooth count.

"Roger, scramble one," Borman replied, using pilot slang for launching a single aircraft.

A cloud of smoke drifts away from Gemini-Titan 6 after its Malfunction Detection System shut down its engine on 12 December 1965.

"The pre-valves are open," Al Bean in the Pad 19 blockhouse told the Gemini 6 crew. The hypergolic propellants were each now one step away from the thrust chamber injectors.

"That's about the best news we've had today," Schirra acknowledged.

"Adios," Bean signed off as the count entered its final 30 seconds.

Jack King counted down the final 10 seconds, called "Ignition!" at 09:54 Cape time and handed the commentary over to Paul Haney, his counterpart in Houston.

[8] The revised mission for Gemini 6 had been redesignated Gemini 6-A.

[9] Alternative spellings Kroner/Kromer/Comer.

Fire, smoke and steam belched out from the flame trench underneath the launch vehicle. Schirra felt the shudder as the engine ignited, saw the thrust-light flicker and then fade, as it was meant to do as the engines built up to full thrust, and saw the clock start as it was meant to do when a plug was pulled from the base of the vehicle as it lifted off. "The clock has started," he duly announced.

In Houston, Chuck Harlan at the Booster console saw the engine thrust build up, and the clock start – and then, to his amazement, the thrust diminished.

In normal circumstances, the Flight Director was the only one in the MOCR authorised to call an abort, but in a launch the pace of events was so rapid that there was no time to make a recommendation to the Flight Director, so Booster had a red toggle switch. He had to verbally call "Abort!" if he decided to throw the switch. It was not a job for the faint of heart. Booster trod a narrow line between calling a false abort and ruining a mission, and failing to call out a valid abort and perhaps losing a crew. Throwing the switch would illuminate a red light in the spacecraft. To guard against a malfunction illuminating the light, Schirra would act only if it was accompanied by a verbal "Abort!" call. Although the clock had started – indicating that the vehicle had lifted at least an inch-and-a-half off the pad, which was what was required to extract the plug – Harlan's instincts told him that it had not moved, so he instead called "No lift-off!"

"We've got a shutdown!" Haney announced to the world that was holding its collective breath. "No lift-off! The engine has shut down!"

Having reached the same conclusion, the Cape began to 'safe' the Titan by isolating its pyrotechnic systems.

Meanwhile, in the spacecraft, Schirra had come to the conclusion that, despite the clock having started, the vehicle was still bolted to the pad. Having ridden on an Atlas, and having witnessed the first attempt to launch Gemini 2 'fizzle', his pilot's instincts told him that the vehicle had not moved. In the simulator, when the regime was to 'prime' a crew to abort, he surely would have yanked on the 'D'-ring, but in real life, trusting 'the seat of his pants', he refrained. The irony was that he had been specifically assured that the clock *could not* start until the vehicle had left the pad. Nevertheless, even as Schirra weighed the conflicting cues his heart rate rose to only 98 beats per minute.

"Oh, shucks!" muttered Stafford on the intercom as his heart rate shot up to 120 beats per minute.[10] By the rules, whether or not to eject was the Command Pilot's decision so he braced himself and waited.

Deke Slayton, who was in the blockhouse with Bean, expected to see a blast of smoke as the crew 'punched out'. Kenneth Hecht, head of the Gemini Escape, Landing and Recovery Office, was also surprised that the crew stayed in the vehicle. However, he was well aware that unless it was clearly a matter of risking *possible* death in order to escape *certain* death, pilots were reluctant to endure the 20-*g* impulse as the seats left the spacecraft. Besides, on one test of the ejection system the doors

---

[10] Or at least that's the NASA version of his remark. Stafford was renowned for his colourful language, and his observation may well have been somewhat stronger.

had failed to release and the suited mannequins had been shredded as the seats had punched through. Furthermore, the mock-up spacecraft had been filled with inert nitrogen, and because Spacecraft 6 contained pure oxygen there was a fair chance that if Schirra chose to eject then he and and Stafford would emerge as flaming candles.

"Programmer reset," came the announcement from the blockhouse. This was good news, because it meant that the Titan's sequencer had been reset, eliminating the danger of some aspect of its logic acting in the mistaken belief that it was airborne.

"Fuel pressure is lowering slowly," Schirra announced. There were meter displays in the cockpit for the fuel pressures in the launch vehicle.

"Roger, Gemini 6, all tanks are venting," Bean acknowledged.

"Our pressures are venting, Flight," the Cape called. "Chris, do you see any problems?"

In Houston, C.C. Williams, sitting alongside Booster and known on the intercom loop as 'Tanks', was monitoring the pressures in the launch vehicle's propellant tanks. "The tanks are venting," he confirmed.

"The bird looks good," confirmed Kraft in Houston.

See put in a call to Gemini 7 to advise them that there had been no lift-off.

"We saw it," acknowledged Borman. "We saw it light up, and we saw it shut down."

Frank Carey, the Martin Company's chief test conductor, told Gemini 6 that it would be 20 minutes before anyone would be able to venture out to the pad.

"That's okay, we're just sitting here breathing," Schirra replied. "Hey, Frank," he added, "I know you guys did your best."

"You can go ahead and stow your 'D' rings," Kroner advised.

"Do you have any diagnosis of our problem yet?" Schirra asked.

It seemed that the electrical umbilical's plug had been shaken out by the vibration as the engine ignited, and when the vehicle's Malfunction Detection System noted this anomalous event it initiated the shutdown. "It looks like one of the tail plugs fell out!"

"Roger."

"Did you get a clock-start?" Kroner asked.

"Affirmative."

"Well that's how it happened."

Schirra was philosophical. "These things happen. It could happen to anyone. No one was hurt."

"Mighty cool head there, Wally," Carey added. "We appreciate it." If Schirra had pulled the 'D'-ring, the ejection would have ruined the spacecraft. By deciding that the vehicle had not left the pad, despite the clock having started, and hence was not in danger of toppling, he had saved the mission.[11]

"It's part of that good training we had," Schirra reflected. Later, he would venture that if he'd punched out he didn't think anyone would have blamed him.

[11] NASA awarded Wally Schirra a Distinguished Service Medal for his decision not to eject. The citation made reference to "his courage and judgment in the face of great personal danger; his calm, precise and immediate perception of the situation that confronted him; and his accurate and critical decisions."

Haney described it as, "An unusual turn of events."

Once things had settled down, Schirra called Kraft. "Tell Frank and Jim that we still want to come up and see them."

Once Gemini 7 established contact with Canary, Jim Fucci brought them up to date. "It looks like a four-day recycle." This window would open at 08:43 local time on Thursday, 16 December.

Once the erector had been raised, Guenter Wendt's team moved in to release Schirra and Stafford. As they rode the elevator down for the second time, many bets were won and lost in the Cape's Press Club.

The Aerospace Corporation had a contract to study each launch, to review procedures. That night, Ben Hohmann invited Schirra and Stafford to sit in on the debrief as he studied the telemetry from the launch vehicle. "Wally, I see something I don't like," said Hohmann. The engine thrust had stalled *before* the MDS issued the shutdown signal. The traces were shown to Colonel John Albert, the Air Force officer in charge of Gemini launches, and he ordered an engine inspection. After further examining the telemetry, Aerojet, the engine's manufacturer, found that the gas generator had lost power, and ordered the engine stripped down *in situ*. After working through the night they found the fault – a dust cover left inside the generator's oxidiser inlet had obstructed the flow. A check of the paperwork found that this had been inserted *months before*, when the engine was being assembled in the factory.

About the only good news was that the Malfunction Detection System that NASA had added to the Titan missile in order to 'man-rate' it had been money well spent. Even so, as Schirra later observed, the fact that there had been a man in the abort 'loop' had saved the spacecraft for another day – an automatic launch escape system like that used on Mercury would have whisked the spacecraft away at the first sign of trouble, even though lift-off had not actually occurred. The result of Aerospace's analysis was the realisation that if the plug had *not* fallen out, the Titan may well have lifted off and fallen back onto the pad seconds later when the engine failed. The plug had fallen out 1.2 seconds after engine ignition. The bolts holding the vehicle to the pad were to fire when the engine attained 77 per cent of its rated thrust. This normally took two seconds. If the MDS noticed the slow start and issued the shutdown prior to this, the bolts would have been inhibited. In the event of the engine failing *after* the vehicle had lifted off, the crew would have had no option but to eject. And when the vehicle fell back onto the pad and exploded it would have severely damaged the only pad set up for Gemini–Titan missions, which would have thrown the remainder of the Program into doubt. Given that the cap had been in the engine all along, this scenario would have occurred if an attempt had been made to launch Gemini 6 in October, so, all in all, the Program had had a very narrow escape.

Flying on

A few hours after the frustration of hearing of Gemini 6's failure to launch, Borman and Lovell took the endurance record that had been set by Gemini 5 a few months earlier. They were now pioneering. As a bonus, See, relaying via Ascension, told them, "you're cleared to choose whatever suit configuration you'd like."

"Hallelujah," replied Lovell, "mine's coming off!" For the first time, both men would be unsuited.[12]

Gemini 7's flight plan, which had been cleared for the rendezvous was rescheduled with experiments. The delay made even more urgent the need to save propellant, and it would be necessary to be selective in the choice of experiments. They were limited to using 3 pounds of propellant per day until the rendezvous. Photography would be feasible only if the target was conveniently located in the window as the spacecraft drifted.

One bit of good news came later in the day when the Cape chose to advance the launch Gemini 6 one day to Wednesday, 15 December. "They're going to attempt to launch 'Six in three days," See reported, "and if they don't make it then, they'll just go the next day." The agency was determined not to miss the opportunity of using Gemini 7 as a rendezvous target.

"We're wishing them all the luck in the world," Lovell assured, "and the friendly target vehicle will be standing by."

Before they retired for the night, Garvin on the *Rose Knot Victor* said, "We're looking forward to Wednesday."

"As a matter of fact," came the response from the spacecraft, "we're almost beginning to look forward to *Saturday*." This was a reference to their return to Earth at the conclusion of the marathon mission.

The next day, See reported the news that even if the plug had not prematurely detached from the base of Gemini 6's launch vehicle, the Malfunction Detection System would have shut down the engine before the bolts blew. "How about that?"

"We'll buy that," Lovell replied.

While they slept, they exceeded Cooper's endurance record of 225 hours and 17 minutes accumulated during his Mercury and Gemini missions. Meanwhile, the gas generator on the launch vehicle was replaced, and the intention to launch on Wednesday confirmed.

"Would you tell Dr LaChance of the Crew Systems Division that his chicken with gravy should be labelled gravy with chicken," Lovell wryly observed at breakfast. CapCom Charlie Bassett promised to pass the word.

Later in the day, a Minuteman missile was launched from Vandenberg Air Force Base on the California coast north of Los Angeles, and Gemini 7 collected radiometric measurements as the warhead re-entered the atmosphere over the island of Kwajalein

[12] In fact, a suitless crew was not new; in 1964 the three-man crew of Voskhod 1 had flown without suits.

in the Marianas some 5,000 nautical miles down the Western Test Range. When the *Range Tracker*, located north of Hawaii, relayed an enquiry as to whether they had seen anything, Borman replied simply, "Bull's eye!" They had secured almost two minutes of data, but the price of the observation was paid in OAMS propellant. As the closing speed had been almost 30,000 miles per hour, the missile had descended through Gemini 7's altitude at a point 650 miles to the northeast, and the closest point of approach was off to the left at a range of 140 miles, so Borman had been hard pressed to manoeuvre to maintain the sensor facing its target. It had looked like a meteor trail viewed obliquely from above. As a result of the observation, the PQI was down to 19 per cent, but the Department of Defense had requested that they try. See warned that they were now at the minimum OAMS required for the rendezvous, and "we'd like you to not use any more fuel today, if you can possibly manage it". By now, the orbit had become slightly elliptical, ranging between 161 and 163 nautical miles, but this was not expected to pose a serious problem during the rendezvous.

As the astronauts slept, at the Cape the Titan was loaded with propellant and the count started once again. After being awakened at 04:00, Schirra and Stafford breakfasted with Al Shepard and set off once again for the pad.

Schirra and Stafford breakfast with Al Shepard before finally setting off to chase Gemini 7.

"We just watched Guenter and his crew load Wally and Tom into the spacecraft," Bassett informed Gemini 7 at T–60 minutes, "and everything seems to be progressing satisfactorily."

The Spirit of 76

"It looks like you'll be having company before long," Keith Kundel advised Gemini 7 on Wednesday, 15 December. As the spacecraft dipped towards Carnarvon's horizon, the hold at T–3 minutes was started so that Carnarvon's tracking could be integrated to calculate the best time to launch.

"We hope you get *two* [spacecraft], on the next pass," Chris Kraft told Kundel.

"So do we!"

Upon being asked for the final crew status report, Schirra replied emphatically, "For the third time, 'Go'!"

Gemini 7 remained radio-silent as it overflew the United States, in order not to distract the troops on the ground.

Al Bean was in the blockhouse as previously, and he counted down the final 10 seconds, confirmed "Ignition!" at 08:37, and then fell silent – as 'Stoney', his job was over.

"Go, you mother!" implored one of the press officers.

"The clock has started," Schirra announced, and this time he felt the vehicle rise off the pad, confirming that they were finally on their way.

The first stage performed flawlessly and fell away.

"Steering looks good from here," Elliot See, in Houston, advised.

"She looks like a dream," Schirra enthused as the second stage refined its trajectory. The Titan had to manoeuvre so that it would insert Gemini 6 into an orbit in the same plane as the target vehicle.

"Attitude errors are all zero," Stafford pointed out on the intercom. The absence of any deviations caused him to wonder whether the pair of attitude indicators were actually really working!

"How about that!" the Command Pilot exclaimed in delight. "Did you read that, Elliot? The attitude errors are *zero*!"

"Roger," See acknowledged matter-of-factly. Half a minute later, after checking the plot board on the main display, he confirmed that they were "right down the line".

"Very good," Schirra agreed.

The second stage continued to fly the nominal trajectory through to SECO, prompting a round of applause in Houston. With this launch, NASA established a new 'first' – four men in orbit at once.

"You can't do any better than that," Deke Slayton radioed up.

"Beautiful!" Stafford observed.

"Okay, we're here." Schirra summed up.

In fact, they had a slight underspeed and the apogee was a few miles lower than planned, but this would be easily corrected at the end of the first revolution. Delighted finally to be in orbit, Schirra and Stafford started to work through their post-insertion checklist as they left the Eastern Test Range behind. Gemini 7, some

The second stage of the Gemini Launch Vehicle is tasked with steering a course designed to intercept the orbit of the Agena target.

1,200 nautical miles ahead, was nearing the Canary Islands.

"Did you copy that insertion?" Kraft asked Jim Fucci, the Canary CapCom.

"Affirmative."

"I think you ought to give that to Spacecraft 7."

"'Seven, Canary," Fucci duly called. "Well, they did it!"

"We didn't get to see the lift-off," Borman reported, "but we saw 'em coming up." Their oblique view of the Cape had been obscured by a bank of cloud, but they had picked up the other vehicle once it had broken through.

Fucci then called the newcomers and told them that radar tracking showed that they had an out-of-plane error of only 0.07 degrees.

"We're in great shape, here," Schirra assured.

"Good to hear that."

"How're the 'Seven boys doing?"

"They're about five minutes ahead of you."

With two spacecraft in orbit, the ground voice network was much busier than usual. The Flight Director's instructions to successive stations were interleaved with conversations with the crew, either by Houston via relay sites, or by remote CapComs. As yet, however, the spacecraft were unable to communicate with each other.

"We're 'Go' for a 4th orbit rendezvous," See advised as Gemini 7 rose over Tananarive. "You can start putting on your suits." The mission rule was that all of the astronauts must be suited for the rendezvous and subsequent proximity operations, in case something untoward happened.

"It looks like you're having some thunderstorms down there," Schirra told Kundel when they reached western Australia. It was his first revolution, and he was playing a weatherman in space. Having been up for over 10 days, the Gemini 7 crew had grown used to the world's dynamic weather system. By now, Houston's Real-Time Computer Complex had calculated the sequence of manoeuvres for the rendezvous, so Kundel read it up, adding that they had a 'Go' for '16-1', a landing window that effectively confirmed that Gemini 6 would be able to complete the planned 24-hour mission.

"How're you doing up there this morning?" Ed Fendell asked when they reached Hawaii.

"It's nice to be up here again," Schirra admitted.

Fendell read up the full data for the first manoeuvre, which was to be made at perigee to raise the apogee slightly, in order to remedy for the undershoot at orbital insertion. Although the manoeuvre was prograde it was to be made BEF and using the forward-firing thrusters to accelerate.

Since Gemini 7's mission was so long, it was 'weighed down' with supplies, and so had a reduced propellant load; Gemini 6 had been loaded with twice as much propellant to provide a safe margin for this first attempt at orbital rendezvous, and all of the manoeuvres were to be performed by Gemini 6.

"Everything looks beautiful," Schirra reported as Gemini 6 flew across the United States on its first revolution.

"We'll contact you after the burn," See promised.

Although the burn over the Gulf of Mexico was imminent, Schirra continued his role as weatherman, noting that, "the Gulf coast is socked in".

Concluding the 24-second burn, Schirra reduced the residuals to zero. "PQI is 87.5 per cent," he told Houston. Being lower, Gemini 6 had been catching up with Gemini 7. By the completion of its first revolution, the range had reduced to 635 nautical miles.

"A red object was trailing us just after the height-adjust," Schirra reported. "It's drifting aft." Having been facing aft for the burn and accelerated, they had drawn away from it. "It definitely had a red colour to it. It's obviously part of the spacecraft."

It had evidently been drifting along with them since pulling clear of the booster. Later, they would decide that the colour was an effect of observing through the filmy coating that had smeared their windows during the ascent. As they flew down the Eastern Test Range, Schirra yawed back around to SEF.

"Everything seems to be coming along fine," See informed Gemini 7 through Ascension. "They got the height-adjust burn over the US." Then as the trailing spacecraft rose above his horizon, he warned, "We're going over to 'Six. We'll talk to you later." Now that Bermuda had provided tracking of Gemini 6's orbit following the first burn, See updated Schirra on the phasing burn at apogee that would raise the perigee by an amount calculated to establish the desired timing. As promised, See returned to catch Gemini 7 before it disappeared. "They'll be making their catch-up manoeuvre just beyond Tananarive. It looks good so far!"

"Very good," Borman acknowledged.

"Are you guys getting your suits on?"

"We've got our suits on." They had taken turns, and it had taken two hours.

Once Gemini 7 had faded, See returned to Gemini 6, "Were you copying?" Schirra had heard only a brief snatch of conversation. "We plan for that to get better," See promised. As they reduced the separation sufficiently to establish a line-of-sight, the two spacecraft would be able to converse directly.

When Carnarvon picked them up Kundel told Gemini 7, "We have nothing for you, this pass. You're looking good on the ground."

"Tell them that the order of the day is still to conserve fuel!" Kraft prompted.

"We'll do our best," Borman promised when the instruction was relayed. He had 16 per cent of his propellant remaining.

"You've been doing real good, so far," Kundel pointed out.

When Gemini 6 appeared, Schirra gave the burn report, "We burned out all the residuals. The PQI is 79 per cent." In Houston, the actual rate of propellant usage was being compared against the nominal predictions. Kundel announced that Houston had telexed him that there was to be an additional height-adjust manoeuvre at the next perigee to tweak the apogee.

On the way to Hawaii, Schirra yawed 90 degrees right and made a brief burn to align the plane of his spacecraft's orbit with that of Gemini 7.

While Gemini 6 was *en route*, Kraft updated Hawaii on the mission so far. "All the burns look very good."

"We're glad to hear that," Fendell replied. "Are you smoking a big long green one?"

"I've been smoking," Kraft admitted, "but not a long green one yet; that will be at T+6 hours!" He was saving his best cigar for the news of a successful rendezvous, which should occur as Gemini 6's clock ticked around to 6 hours at the end of the fourth orbit.

"How about turning your transponder on," Fendell prompted Gemini 7 once it appeared.

"Transponder is on."

"How're you doing?" Fendell asked, turning his attention to the trailing spacecraft.

"Very good," Schirra noted. "We completed the plane change – no residuals. The PQI is 75 per cent."

Fendell then read up the data for the tweak. The one-second firing would add a mere 0.8 foot per second and raise the apogee by a few thousand feet to a level precisely 15 nautical miles below the orbit of the target vehicle, as required for the nominal co-elliptic rendezvous.

"Everything's looking real good on both birds," Fendell told Kraft. Then, to Gemini 6, he asked: "Do you require anything else?"

"We're just watching the Sun come up," Schirra replied. Stafford had actually donned his sunglasses to cut down the glare.

"Your cohort is doing real fine," Fendell reported. Then he switched to Gemini 7 before he lost it. "Why don't you give 'Six a shout?"

"Hello Gemini 6, this is 'Seven. How do you read?"

"It doesn't sound like you got to him," Fendell pointed out when there was no response. "Probably by the next time around you'll be reading him loud and clear."

Once Guaymas had picked them up, See recommended that they switch antennas in an effort to improve the chances of establishing direct radio contact, but to no effect.

Gemini 6, trailing a few minutes behind, performed the tweak burn approaching the west coast. "It looked good here, Flight," Guaymas reported, after monitoring the telemetry from Gemini 6.

If the spaceborne radar met its specifications and acquired the target's transponder at a range of 200 nautical miles, then by the time Gemini 6 started across the Atlantic on its third revolution it should be tracking its own target – the first time that one spacecraft had tracked another by radar.

"Where's Jim? Out to lunch?" See wondered when Borman replied to his request for an update on Gemini 7's fuel cell status; the indicators were on Lovell's side of the spacecraft.

"Not exactly," Borman replied, "but he's busy!"

"We copy," See assured. Lovell was in the process of 'managing' his waste with a 'bluc bag'.

"Right in the middle of the rendezvous!" Lovell wryly added.

"Does Gemini 7 have their transponder on?" Schirra called Houston. They were close to the nominal acquisition range.

"Affirmative," See assured.

"And they're BEF, is that correct?" With Gemini 6 playing catch-up, Gemini 7 had to be flying backwards in order to aim its transponder at its pursuer.

"We're getting lights on the radar," Schirra reported a few seconds after See confirmed Gemini 7's attitude, "but no lock-on yet."

Just before Gemini 6 followed Gemini 7 beyond the Eastern Test Range, See read up the data for the co-elliptic burn at apogee to circularise 15 nautical miles below the target's orbit.

"We have what looks like a positive lock-on," Stafford reported through Ascension.

"Roger," See acknowledged.

At this point, direct communication was established between the two craft. "Gemini 6," Borman called, "how do you read?"

"Loud and clear, fella!" Schirra replied.

"We hear you loud and clear also," Borman finished the handshake.

"We'll see you soon," Schirra promised.

The co-elliptic burn was made just after Tananarive. The specification included a slight downward component, and instead of holding the nose depressed throughout the 35-second firing, Schirra burned the horizontal component SEF and then dipped the nose 2 degrees to add the downward component immediately prior to taking out the residuals.

Meanwhile, Kraft was talking to Carnarvon on the network. "Tell Spacecraft 7 that the cut-off for station-keeping is 11 per cent, and under no circumstances are they to use their reserve tank. Also, we want a PQI from Spacecraft 6."

"Did I understand, P-Q-I?" Kundel queried.

"Propellant Quantity Indicated," Kraft confirmed. Then as an afterthought he rebuked, "That's flight control talk!"

"Yessir!"

"Sorry about that," Kraft apologised.

Once Gemini 7 was in range Borman reported, "We can't tell if our acquisition lights are working or not – Wally can't see them." In addition to the radar transponder, his spacecraft was fitted with the Agena's blinking lights to enable Schirra to observe his target against the stars during the terminal phase, which would be flown in darkness, but as they were still 155 nautical miles out it was not really surprising that he could not yet see his quarry.

Following up on the Flight Director's instructions, Kundel asked about propellant, and Schirra reported that they had 68 per cent, which was excellent.

Over Hawaii, Stafford came on the air and announced, "Hawaii, standing by."

"Who'm I talking to!?" Fendell prompted.

"Gemini 6."

"Go ahead."

"Standing by!" Stafford repeated.

As they closed within 100 miles, slowly catching up with their target, Stafford started the computer in a cycle of sampling the radar every 100 seconds to measure the range and range-rate, and he began to plot the range against the elevation (measured by the spacecraft's pitch angle) on a graph whose axes formed target-centred polar coordinates and confirmed that they were maintaining a constant differential of 15 nautical miles.

"The radar needles are a little spongier than we're used to seeing," Schirra told Houston, upon establishing contact through Guaymas. He could aim the spacecraft's nose towards the target by measuring the strength of the radar return but the reading was fluctuating. "It's plus or minus one or two degrees."

Before Gemini 6 left the United States behind, See read up Houston's calculation for the TPI burn, based on tracking by Carnarvon and Hawaii, as a backup to Stafford's onboard determination.

The Earth's rotation took the spacecraft's track west of Canary, so the *Rose Knot*

Victor, which was stationed in the Atlantic off the bulge of South America, acquired Gemini 6 for the first time.

With Gemini 6's nose pitched up at 19 degrees facing the target, the Sun was behind the adapter module and the windows were in shadow as orbital sunset loomed.

"I have a lighted target at about 12 o'clock," Schirra noted. "It could be a star, but it might be 'Seven. We'll check it out."

Half a minute later, Stafford told Schirra excitedly on the intercom. "Hey! I think we've got them. That's 'Seven, Wally!"

"Yes?"

"It's either Sirius, or 'Seven," Stafford decided. Sirius was not only the most prominent star in the sky, at this point in the chase it would be close to their target in the sky.

"We either have Sirius, or 'Seven," Schirra told See as Ascension Island picked them up. "It's right in the middle of the reticle." His optical reticle was boresighted to the radar, which was indicating a range of 48.5 nautical miles. "The radar's locked-on." Four minutes later, he was sure. "I now see the Belt of Orion, so Sirius is below our nose. And *that* isn't a star that I recognise." It was a spacecraft in a higher orbit reflecting the Sun. It moved against the stars as the catch-up phase progressed. After another two minutes, Schirra reported, "The target is in line through the centre of Orion's Belt."

They had designed the mission patch featuring the 'Northern Six' stars, which navigators had used for centuries. As Stafford reflected after the mission, "We were up there aiming for the rendezvous, and when we first saw our rendezvous vehicle glittering in the reflected light of the sunset, it was right between Sirius and the twins, Castor and Pollux, the two brightest stars in the constellation of Gemini, exactly where we had placed it on the patch."

The patch for Gemini 6 depicts the stars against which the target vehicle would be viewed in the terminal phase of the rendezvous.

Considering that their target was catching the last rays of the Sun so brilliantly, Schirra was amazed that he had not seen it earlier. "This is better than in the simulator!" he opined. "We've got you, Frank!"

"Roger," Borman acknowledged.

As Gemini 6's lower orbit brought it in below Gemini 7, Schirra kept the nose facing the target and Stafford monitored the increasing pitch. Meanwhile, Borman was pitching his own vehicle down in order to maintain its transponder facing the inbound spacecraft. As Gemini 6 pitched up towards 25 degrees, its computer processed three measurements in a 'closed loop' algorithm to calculate when the terminal phase should be initiated. The information supplied by

Houston was to provide a backup in case the onboard determination was foiled. As they had made the co-elliptic burn slightly early, and been two miles short of the best manoeuvre point, Stafford knew that TPI would have to be delayed beyond Houston's predicted time to correct for this. In fact, the key parameter was not the timing, but the geometry; they had to make the burn when the angle was 27 degrees, at which time the range would be 33 miles. The polar plot continued to chart their co-ellipticity as nominal, so he was sure of their trajectory.

When the target flew into the Earth's shadow it disappeared. As Gemini 6 followed it in, Schirra discovered that he could not see its acquisition lights. With the radar needles giving a 'spongy' response, he *needed* to see the target, but his eyes had yet to adapt to the onset of darkness. "Do *you* have them in sight?" he asked Stafford, whose night vision was probably better because his attention had been inside the dimly lit cabin.

"No," Stafford replied.

Meanwhile, in Houston, Kraft was congratulating his team for directing Gemini 6 to the initiation point for the terminal phase. The interception would be up to Schirra and Stafford, flying a spacecraft with its own radar and a state-of-the-art computer.

Schirra used the radar to hold his spacecraft facing its target as best it allowed, and when the pitch angle was right he initiated the transfer by firing the aft-thrusters for 41 seconds to speed up by 33 feet per second. Three minutes later, having reduced the illumination on his reticle to its minimum, he finally saw a speck of light close to the centre. "I've got a real dim light out there." Logic dictated that this was Gemini 7, but because it was no brighter than a fifth magnitude star he was not sure.[13]

"We'll blink it a couple of times," Lovell announced. He switched the acquisition lights off. "It's off, now."

"Okay, that was it," Schirra agreed.

"Coming on, now," Lovell resumed.

Now that Schirra had positively identified his target, he momentarily dropped the nose down to the horizon to verify the calibration of the platform. It was now crucial that he manoeuvre to maintain the target fixed with respect to Castor and Pollux and hence pursue a line-of-sight trajectory that would result in an interception. Although the approach was 'straight in' from Schirra's point of view, in terms of the relative motions of the two spacecraft around the Earth the terminal phase subtended a central angle of 130 degrees and traced the shape of a fish hook extending from the TPI point on Gemini 6's co-elliptic orbit to the interception point on the higher orbit.

"It's really bright now," Stafford ventured 16 minutes into the interception, by which time Gemini 6 was pitched up at almost 70 degrees.

"We're sneaking up to meet you, Frank!" Schirra called to Borman.

When the *Coastal Sentry Quebec* picked up the spacecraft, Lewis listened in

---

[13] "We were somewhat horrified how dim they were," Schirra would later reflect, and he recommended that the power of the Agena's acquisition lights be increased before anyone chased one down.

silence in order not to distract the crews during this critical phase of the rendezvous because, as Kraft had said, they were now on their own. Stafford's running commentary of the increasing pitch angle and the radar's range and range-rate enabled everyone to keep track of their progress. For most of the mission, Stafford's attention had been firmly inside the spacecraft. Since the co-elliptic burn, he had been logging the computer's results in his notebook.

"One [nautical] mile, Wally," Stafford noted, "at 127 degrees."

"Holding steady on the ball," Schirra pointed out. The increasing pitch had eased off. It would soon be time to start the braking phase which, if they did it correctly, would result in them drawing up alongside their target with zero relative motion.

"Look at that light!" Stafford exclaimed as Gemini 7 popped into sunlight. The transition was so sudden and so stark that it dazzled his dark-adapted eyes.

"How about that!" Schirra exclaimed.

Just as Gemini 6, trailing Gemini 7, had entered the Earth's shadow with the Sun behind its adapter, having pitched up to maintain its nose facing the target, the Sun was still behind its adapter on emerging into daylight slightly below and ahead of its target.

"Stand by to brake," Stafford warned 31 minutes after initiating the transfer, as the range approached half a mile.

"Ready to brake," Schirra confirmed.

"Brake at 0.48 miles."

The burn with the forward-firing thrusters reduced the range-rate by 42.5 feet per second. Unfortunately, the thruster plume caught the Sun, and dazzled Schirra and for a while he lost sight of the fainter stars.

"1,200 feet, at 33 feet per second," Stafford reported, a minute and a half after the burn. "We're in good shape."

By this point, the reflection from Gemini 7's white adapter was so bright that Schirra lost the stars completely. "They look like they're on fire!" he complained. But by now the range was down to 300 feet and it was evident that the braking was going well, and so he was not concerned.

"We've got company," Lovell informed Borman on the intercom. They had watched in fascination as the other vehicle rose towards them. The plumes from the braking thrusters were spectacular.

"Having fun!?" Borman called over the radio.

"Hello there!" Schirra greeted.

As Lovell recalled, "It was just becoming light. We were face down. And coming out of the murky blackness below was this little pinpoint of light. The Sun was not illuminating the ground yet, but on the adapter of Gemini 6 we could see sunlight glinting." They had not been able to see the visitor earlier in the transfer because Schirra, not wishing his docking light to reflect off the nose of his vehicle and impair his night-vision, had not lit it. "And as it came closer and closer, like it was on rails, it became a half-moon shape. At about half a mile, we could see the thrusters firing, like water from a hose – and just in front of us, it stopped!"

"Holding steady at 120 feet, Wally," Stafford said, continuing his running commentary. "You can turn off your light, Frank, if you want to save power." There was no need for the lights in daylight.

The final phase of the rendezvous had taken place over the Central Pacific, northwest of Guam. The *Range Tracker* was stationed on the International Date Line to provide a relay at this crucial time. The tension in Houston was electric. As the two vehicles had dipped below the *Coastal Sentry Quebec's* horizon everything looked promising, but Gemini 6 had not yet started to brake, and there was a distinct possibility that it would shoot straight past. If so, it would not have been able simply to turn around and fly back, the rendezvous would have been missed and orbital dynamics would have drawn them apart.

"Gemini 6, Houston," Elliot See called when the *Range Tracker* reported that it had the spacecraft's beacon. "Standing by."

"Go ahead," Stafford replied.

"We're interested in your status!"

"What range are we, Tom?" Schirra asked on intercom.

"120 feet, steady."

"We're station-keeping at 120 feet," Schirra radioed.

With this news, the flight controllers cheered and mounted tiny American flags on top of their consoles. Jerry Bostick, the Flight Dynamics Officer who had done more than anybody to steer Gemini 6 to the TPI point, had toured the undertakers of Clear Lake and purchased their entire stock of the pennants used to adorn the vehicles of military funeral processions.

"There must be fully 40 flags in this room," Public Affairs Officer Paul Haney continued his real-time commentary. "Everyone's standing. The viewing room is jammed with people!"

The quality of the link through the *Range Tracker* was poor, so See said that they would wait until Hawaii for further details.

When Hawaii picked them up, Schirra confirmed their status, "We're in formation with 'Seven, and everything is 'Go' here."

"Congratulations!" See enthused. "Excellent!"

"Thank you," Schirra replied. "It was a lot of fun." He continued to pitch over until he was lined up with the horizon, inverted, and then rolled 180 degrees in order to fly upright.

Kraft delightedly lit up his celebratory cigar. He was joined by Gilruth and Joe Shea, the Apollo Program Manager. The rendezvous went a long way to mitigating the risk in the Apollo 'mission mode'. Confounding the skeptics, Gemini had achieved rendezvous at the first attempt. In a sense, the door to the Moon had just been thrown open.

As Gemini 6 had completed the rendezvous, Stafford had observed that there was "a big string hanging" from the rear of Gemini 7. The occasional shadows that had been cast across that spacecraft's nose had led its crew to suspect that there was something dangling off their adapter module. In fact, Gemini 6 had one too. The need for Gemini 7 to minimise propellant usage restricted the extent to which Borman could manoeuvre, but Gemini 6 had plenty of fuel left – the PQI was 50 per cent –

and Schirra flew a loop around Gemini 7 to inspect its condition. Apart from when Ed White had spacewalked on Gemini 4, this was the first time that anyone had been able to inspect a Gemini spacecraft in space, and White had not seen streamers on the rear of his vehicle. The strands were a few inches wide and 10 to 15 feet long. "Well, Frank, it looks like it comes off the separation plane," Schirra reported.

"The separation plane from the booster, right?"

"Affirmative. It might be the fibreglass."

Gemini 6 made a fly-around inspection to document
the condition of Gemini 7, particularly the 'streamers' at the rear.

"That's exactly where you have one, too," Borman observed. "It really thrashed around when you were firing your thrusters." It was later decided that the debris was the bundles of primer cord from the booster's separation plane.

"'Six and 'Seven, Hawaii," Fendell called to remind the two vehicles that they were still within range, "we'll be standing by, if you have anything for us."

"There just seems to be a lot of traffic up here, that's all," Stafford pointed out.

"Call a policeman," Borman retorted.

"What's your range?" Fendell asked.

"20 feet," Schirra replied.

"We did it!" Fendell congratulated Kraft moments after the two spacecraft flew beyond his range.

Prior to launch, Schirra had emphasised to reporters that what the Soviet Union had twice done by directly inserting two vehicles into similar orbits a few miles apart did *not* constitute rendezvous. Gravity drew them apart. The Vostoks had been incapable of manoeuvring and so had been unable to close the gap.[14] Rendezvous, Schirra had insisted, meant flying in the *same* orbit. Within about a hundred feet, differential gravity effects were negligible, and so Schirra was able to 'park' Gemini 6 just in front of Gemini 7's nose and, because they were flying in near-vacuum, there were no forces acting to disturb this relationship. With one inch per second velocity increments, the Gemini spacecraft proved to be extremely controllable, prompting Schirra to opine that lining up to dock with the Agena would be straightforward. "It'll probably be easier than in the simulator."

An in-plane fly-around had been planned in the first daylight pass, but Schirra postponed this so as to practice manoeuvring in close proximity instead. His 'cycle' of the instruments was slower than he would have preferred because it took time for his eyes to adjust to read the instruments in the cabin after being dazzled by the glare from Gemini 7's white-painted adapter.

"Right now, we're SEF, and 'Seven is BEF," Stafford announced when Guaymas picked them up, indicating that they were flying nose-to-nose, "and I'd say we're 10 feet apart." As Schirra continued to manoeuvre to inspect his companion, Lovell observed, "I can see your lateral thrusters firing out about 40 feet." It had been agreed that the vehicles would not be permitted to come into physical contact, because it was possible that they had picked up an electrostatic charge by flying through the ionosphere, the discharge of which might damage the delicate electronic systems.[15] The closest that Schirra approached was about 12 inches. When they flew out across the Atlantic in orbital darkness, he held Gemini 7 in the beam of his docking light and concluded that, contrary to expectation, station-keeping was easier in the dark.

Meanwhile, in Houston, in his ongoing commentary, Haney reflected upon the mood in the MOCR. "One can almost sense the feeling of achievement evident in the room. You'll have to go back to the Shepard flight – at least, in my memory – for a time when all of the flight controllers stood at their consoles like the moment when this rendezvous occurred and we got that first report."

Back in October when the dual mission plan had been hatched, Walter Burke had offered to develop an inflatable collar for the rear of Gemini 7's adapter to enable Gemini 6 to assess lining up to dock, but Borman had steadfastly refused to sanction such a folly. Like Schirra, he preferred to focus on the primary objective. At that stage in the Program, rendezvous was the major 'unknown'. If they managed to achieve this and then an impromptu collar foiled a docking rehearsal, the mission would be

---

[14] The Soviet Union's communiques regarding these 'double' Vostok flights had been carefully worded to imply that the spacecraft had manoeuvred following insertion to reduce the initial separation, but in fact the reduction was a result of orbital dynamics, and the range opened up again and thereafter increased continuously. It has to be born in mind, however, that NASA knew little about contemporary Soviet capabilities. Pavel Popovich, one of the cosmonauts involved, was quoted by *Izvestia* on 21 December 1965 saying that Vostok 3 and 4 had come to within 3 miles of one another so "the American experiment of an orbital rendezvous repeated to some degree what we did", but noted that "techniques have advanced a great deal". On being informed of this the next day, Schirra was dismissive. "If anybody think's they've pulled off a rendezvous at 3 miles, have fun! This is when we *started* doing our work. I don't think a rendezvous is over until you are stopped – completely stopped with no relative motion between the two vehicles – at a range of approximately 120 feet. That's rendezvous!"

[15] The Agena's docking apparatus incorporated 'whiskers' designed to bleed off any electrostatic charge before the Gemini spacecraft's nose touched the wall of the docking collar.

dismissed as a failure. The act of docking, Borman had been adamant, could await the fix to the Agena's engine. In retrospect, even although a docking was not possible, it was far more rewarding that the first rendezvous was made by two manned spacecraft instead of a manned vehicle chasing down an unmanned target. The pictures of the spacecraft manoeuvring a few feet apart became instant classics, and were a public relations boon for the agency.

"How's the food supply holding out?" Schirra asked the crew of the marathon mission.

"Oh, we're in good shape," Borman assured.

"It's holding out," Lovell added, "but it's the same thing from day to day."

In fact, as Borman noted in the post-flight debrief: "I guess the people didn't realise we were going to operate on a regular day, and try to eat breakfast, lunch and dinner. The meals weren't prepared that way, and often times we had shrimp cocktail and peas for breakfast!"

Borman and Lovell knew that if it had not been for the prospect of entertaining visitors, the final few days of their 14-day endurance mission would have stretched their stoicism to the limit.

"Could you give us a report on your night-time station-keeping?" Bill Garvin asked when they rose above the *Rose Knot Victor's* horizon.

"No trouble at all," Stafford assured. Schirra had drawn back to about 20 feet, and they were keeping track of one another via the lights.

In the darkness, they could also see through one anothers' windows, into the illuminated cabins. "Can you see Frank's beard, Wally?" Lovell asked.

"Yes, I see yours better, right now."

"For once, we're in style," Borman wryly observed, making reference to the hippy style that was in vogue across most of America – with the notable exception of NASA facilities.

"How's the visibility through *our* windows?" Stafford prompted. The panes were coated by a smear that was more opaque than usual.

"It's pretty bad," Lovell confirmed. "We noticed at sunset that we could barely see in your windows."

Immediately after sunrise, Schirra began the in-plane fly-around. "Just ignore us," he told Borman.

"Do you have the feeling that somebody's back behind us?" Lovell asked Borman, when Gemini 6 disappeared from sight.

Stafford then flew the out-of-plane loop in 10 minutes – half of the time it took in-plane. Their conclusion was that it was much easier to manoeuvre out-of-plane because the horizon remained visible as a reference throughout.

With the day's objective achieved, and the two crews settling down for the night, Kraft handed his console over to Gene Kranz and then strolled over to the Press Room, where he received a champagne toast from the massed reporters.

After three revolutions in close proximity, Gemini 6 made a 15-second retrograde burn and – by the paradox of orbital dynamics – drew ahead of Gemini 7. As the range opened, Stafford used an experimental sextant to track the target. This had been added to the flight plan at the last minute to assess optical tracking for Apollo.[16] While the instrument worked, his conclusion was that it was too bulky for routine use. Having opted not to eat during the rendezvous or proximity operations, Schirra and Stafford were ravenous, so they devoured a meal and then retired.

The following morning, on the next-to-final pass over the United States, Schirra stunned the flight controllers with an urgent call, "We have an object – it looks like a satellite – going from north to south, up in polar orbit. He's in a very low trajectory ... it looks like he may be going to re-enter pretty soon." This prompted some concern on the ground, because the only object likely to fly low over the pole and re-enter was an intercontinental ballistic missile. "I see a command module," he continued, "and eight smaller modules in front. The pilot of the command module is wearing a red suit ... he looks like he's trying to signal us!" He now gave a harmonica rendition of 'Jingle Bells' and Stafford joined in, shaking a set of miniature bells, possibly the first musical instruments in space. Christmas was only 9 days away and Schirra had wanted to play a 'gotcha' on the controllers; his lead-in was so matter-of-fact that – just for a moment – he had had everyone fooled.

"We can also see the object," Lovell chimed in.

"That was 'live'," Schirra noted, referring to the musical accompaniment, "not taped."

"You're too much, 'Six," See dismissed.

Controlled descent

A few hours later, See counted down to retrofire via Canton Island and Schirra reported that all four rockets had fired. Gemini 7, above and in trail of Gemini 6, was to have filmed this, but several minutes beforehand Borman had realised that two of his yaw thrusters were no longer effective, and he had been unable to line up to take the shot.[17]

"This bird's been a beauty all the way," Schirra told Fendell in Hawaii after the de-orbit burn. Then he switched to HF and made a final call to Gemini 7. "Really a good job, Frank and Jim. We'll see you on the beach."

"Okay, Wally," Borman replied. He and Lovell still had two days to endure before they would be able to follow suit.

Given the less than perfect descents of the earlier missions, when it was announced that they were to fly a dual mission Schirra had placed a bet with Borman as to which spacecraft would make the more accurate descent.[18]

After jettisoning the retro section of the adapter module, Schirra rotated the spacecraft 'heads down' to enable him to monitor the horizon during the early stage

---

[16] The sextant had not been part of the flight plan for the original Agena mission.

[17] When commanded to fire, the thrusters were venting unburned propellant. Post-flight analysis revealed that the laminate in the thrust chambers was *not* of the specification introduced specifically to rectify this burnout fault, which had revealed itself on previous missions.

[18] In '*Schirra's Space*', Schirra later wrote that no matter how much he rehearsed his Mercury re-entry in the simulator he had never been able to hit the target. It was only when the engineers realised that the calculations were flawed that he realised that he had been consistently splashing down within several miles of the target, so he was particularly optimistic about winning his bet with Borman.

After splashing down Wally Schirra temporarily opened his hatch (top), but in contrast to his predecessors he refused a helicopter pickup, and did not disembark until the capsule was on the deck of the *USS Wasp*.

of re-entry. At 60 miles altitude he banked 55 degrees left and held this angle until down to 50 miles, at which time he began to follow the computer's steering cues.

The capsule splashed down just over 7 miles from the target point in the prime recovery zone some 500 miles southwest of Bermuda. Because this was within sight of the *USS Wasp*, for the first time it was possible to relay television of a recovery operation through the Early Bird satellite which had been placed into geostationary orbit above the Atlantic earlier in the year. Once the flotation collar was on, Schirra and Stafford opened the hatches and chatted with the swimmers. Being so close to the carrier, they declined the helicopter pick up, opting instead to ride the hoist. With the capsule safely on the carrier's deck, Schirra and Stafford jumped down onto the red carpet. The ship's island structure was adorned by a banner that read: '*Spirit of 76*'.

Gemini 7's final days

As soon as Gemini 6 had withdrawn, Borman and Lovell eagerly took off their pressure suits to fly the rest of the mission in their longjohns. Not long after, two of the six 'stacks' within the fuel cell system ceased to produce power. Just before they retired, the astronauts achieved another milestone by exceeding the durations of all the previous Gemini missions combined. While they slept, Hodge's flight controllers reviewed the power situation and in light of the facts that the PQI was now down to 8 per cent, two thrusters had failed, and the experiments were already curtailed, it was decided that there was no cause for concern. "As things look now," Hodge said, "I see no reason why we can't complete the 14-day mission." As the crew ate their breakfast they were given the go-ahead for '207–1', which was the full duration. If the fuel cells were to fail, the 10 hours of battery life would enable them to hold out for the prime recovery zone.[19]

---

[19] It seemed that the venting water boiler marked the onset of the fuel cell problem. Despite this problem, the fuel cell system was deemed sufficiently reliable for use on the remaining missions, each of which would last only three or four days.

Just as the extended build up to the rendezvous had given Borman and Lovell something to look forward to, when it was over their spirits declined and, facing power and propellant limitations, they continued their routine of biomedical measurements and counted the orbits in much the same mindset as Cooper and Conrad towards the end of their mission. Recalling Conrad's advice, they made a start on their novels.

As they entered their final 24-hours, Borman wanted to know how accurately Gemini 6 had splashed down as he had "a lot riding" on making a more accurate re-entry than Schirra. Garvin at Carnarvon relayed the Flight Director's verdict on the bet: "you're going to have trouble collecting".

Later, See informed them that the 135 feet of movie footage shot by Gemini 6 during the rendezvous had just been shown on one of the screens in the control room and had raised an ovation from the flight controllers on duty and the packed VIP gallery. "You look just great up there."

"Can you see our big long piece of strap?" Borman enquired.

"Sure do."

After a few minutes See was back. "Your pictures are on all three TV networks."

"And here we are in our underwear," said Borman lightheartedly.

By this point, trash was becoming a problem. They had not been able to stuff as much meal packaging behind the seats as hoped, and were now collecting it in bags that would be pushed into the footwells for re-entry. "I hope you can still see out," See teased.

"We're really in pretty good shape," Borman assured.

The sleep cycle was adjusted to favour the final day, to ensure that they would be well-rested for their return to Earth.

On their final morning, with six hours to go to retrofire, Borman and Lovell wriggled into their suits. With the PQI down to 5 per cent and two thrusters out, Borman reported that if the OAMS proved ineffective he would activate the RCS early so as to stabilise the vehicle. On the HF uplink, the ground stations played "I'll be Home for Christmas". After 14 days they were eager to return to Earth, even though doing so would mean enduring the crushing load of re-entry. Borman requested a helicopter pick-up – he had no intention of bobbing on the ocean swell waiting for the carrier to steam alongside. And of course, the medics were eager to get them onboard as soon as possible to observe how their bodies reacted to gravity after such a lengthy period in weightlessness, and to monitor their recovery – if the recovery was rapid, then every minute wasted sitting in the capsule on the ocean would be hard-won data irretrievably lost. As Gemini 7 flew its final revolution, the CapComs at each tracking station in turn wished them well, signed off and played 'Going Back to Houston'.

The de-orbit manoeuvre was made in darkness over Canton Island. One worry was that after a 14-day 'cold soak' the solid rockets might misfire, but they all fired – although again there was a slight pause before the fourth ignited. After so long in weightlessness, Borman and Lovell were amazed by the force of the rockets. As they flew over Hawaii, Fendell said that they would would not have an illuminated horizon to check their instruments until they were down to 350,000 feet. "So long, Hawaii," Borman signed off. The 4 g re-entry loading "felt like a ton". They splashed down at

09:05 Cape time on Saturday, 18 December. During their record-breaking mission of 330 hours and 35 minutes they had circled the Earth a total of 206 times and flown 5 million miles – while this made them the most-travelled humans in history, such statistics did not really mean very much any more.

Although, to Borman's delight the computer was able to trim half a mile off Gemini 6's recently set record, Schirra jokingly attributed Borman's success to the excellent advice that he had supplied after his own descent! As it was hot in the capsule, Borman suggested that they doff their suits, but they discovered that they did not have the energy – the gravity was overwhelming. In an effort to compensate, they increased the oxygen flow and tried to relax while the swimmers attached the flotation collar.

Exhausted but delighted Jim Lovell and Frank Borman arrive on the deck of the *USS Wasp*.

Between them, Geminis 6 and 7 had shown that the computer could accurately measure an atmospheric entry trajectory and provide cues to enable the pilot to steer to within visible range of a recovery force. If the next few missions could repeat this, the computer would be allowed to take over and bring the final missions home 'closed loop'.

The objective of the Gemini 7 mission was to determine how selectively and at what rate the human body reacted to the space environment, and to gravity on return to Earth. There was concern that they might have deteriorated to the extent that return would be debilitating. Borman and Lovell rode the 'horse collar' hoist up to the helicopter, the forces of the ascent further abusing their space-adapted bodies. Within half an hour of splashing down they were on the deck of the *USS Wasp*. Their arrival was broadcast 'live'. Although they were tired, it was apparent that they were very pleased with themselves. They were unshaven and as pale as ghosts, but they wore the broad grins of a job well done and were clearly not in physical distress. Shunning assistance, they made their way slowly down the red carpet. Lovell later noted that he had had to consciously work his legs in order to walk – left, right, left – and compared his stooped, knees-apart gait to how he would have walked if he had "filled his pants".

Assessment

After their eight-day flight, Cooper and Conrad were tired, weak and dehydrated, and it had been expected that Borman and Lovell would be even worse, but they proved to be in *better* physiological condition. "Apparently, there had been enough time for an *adaptive* phenomenon to take place", Dr Berry observed. This was encouraging, as it suggested that the undesirable effects tailed off and that the body was readily able to recover. Borman had lost 10 pounds of body mass and Lovell had lost 6 pounds. Lovell's pulsating thigh cuff had had no measurable effect. Borman's EEG waves showed that he had slept fitfully for only a few hours on the first night (which turned out to be the worst night from the standpoint of sleep) but this experiment had been curtailed when he accidentally disconnected the sensors. Nevertheless, he had slept well thereafter. "Our watches were set on Houston time," he said. "We had a regular work day, three meals a day, and then at night we went to bed. We put up light filters in the windows and didn't look out – to us that was night time." As measured in terms of heart rate induced by exercising in space using a bungee cord, their stamina had not decreased appreciably. The bone density data was surprising: at 3 per cent, the bone loss was one-third of that suffered by Cooper and Conrad, even though the mission was much longer. The healthy state of the Gemini 7 crew was attributed in part to the more comfortable G5C suits, which could be removed, the fuel cell fault that had made Gemini 5's cabin chilly, and the increased psychological stress of nursing the ailing Gemini 5 spacecraft. "If we had had the suits that had been used prior to our flight – and not been able to remove them – I think our physiological effects would have been tremendous," Borman pointed out, "and it would have been a matter of survival rather than a matter of operating efficiently." A major part in their success, of course, had been Borman's sheer determination to accomplish the assigned mission. He likened the flight to spending "a fortnight in the front of a Volkswagen", and on reflection he added, "it seemed more like six weeks than a fortnight". To Lovell, it was "like spending 14 days in a men's room". Early on, Lovell had lost his toothbrush and for the rest of the flight they had shared. Despite constantly knowing where they were on the flight plan, they had etched marks on the control panel to count the days. Their mission was longer than all of the Soviet missions added together!

In terms of Press coverage, Gemini 7 marked a transition. The shorter previous missions had been given blanket coverage on TV and radio. With so little happening on Gemini 7, the coverage slipped to a 'round up' in scheduled bulletins. In truth, of course, the drama of the dual mission was at Pad 19 in the frantic effort to prepare Gemini 6. While the role of target vehicle played by Gemini 7 was unglamorous, this improvisation had advanced three of the Program's objectives. Demonstrating rendezvous and station-keeping confirmed that Apollo would be able to utilise the Lunar Orbit Rendezvous mission mode. The controlled re-entries marked a significant step towards verifying that an Apollo returning from the Moon would be able to use the two-stage 'skipping' re-entry. And the 14-day flight confirmed that astronauts would be able to function in space for long enough to mount a lunar surface expedition.[20,21]

"It has been a fabulous year for manned spaceflight", reflected Gilruth once Gemini 7 had splashed down. Although *rendezvous* had been shown to be feasible, the task of docking was outstanding, and this could not be attempted until the fault in the Agena's engine was traced and rectified.

After the post-mission debrief, which lasted several weeks, Slayton assigned Grissom to command the first Apollo mission, with Schirra in backup; Stafford was assigned as backup commander to Gemini 9; and Young was rotated to command Gemini 10.

---

[20] On the other hand, it had yet to be determined if the space radiation and the micrometeoroid rates would jeopardise such a mission.

[21] Biomedical research would not be granted such a high profile on an American mission until the flight of the Skylab space station.

Chapter 6

First Docking
Gemini 8

Fixing the Agena

While magnificent improvisation following the loss of the first Agena target vehicle had resulted in Gemini 6 and Gemini 7 making the first rendezvous by two manned spacecraft, the act of docking had yet to be achieved. Neil Armstrong and Dave Scott were eager to try on Gemini 8, but they had to wait while the fault in the Agena's Primary Propulsion System was investigated.

The Agena had a reputation for reliability, but the GATV was not the standard version, and it was a fair bet that the fault was caused by the modification of the engine ordered by NASA to provide a multiple restart capability. Identifying the precise cause of the engine's explosion from the brief moment of telemetry available was a daunting task. Two days after GATV-5002 blew up, an Agena Review Board was formed under the chairmanship of Robert Gilruth, and it took four months of seven-day-weeks and round-the-clock shifts to identify and fix the fault. By 1 November 1965, it was clear that a combustion instability had been caused by the fact that fuel was injected into the chamber before the oxidiser, and this had resulted in a 'hard start' in which the pressure transient had blown the engine apart, which had destroyed the vehicle. In the Air Force's Agena-D, which could be restarted only once, the flow of oxidiser was initiated prior to feeding fuel because tests had established that this improved the engine's performance. However, because NASA intended to use the Agena for extensive orbital manoeuvring, it had rejected this 'waste' of oxidiser in firing up the engine and had ordered that the fuel and oxidiser be injected simultaneously. Despite the extensive ground tests with GATV-5001, something had evidently delayed the arrival of the oxidiser during the ignition of GATV-5002's engine, with the result that the fuel-rich mixture in the chamber had burned explosively.

To provide a contingency against the repairs to the Agena dragging on, John Yardley of McDonnell suggested that a Re-entry Control System from a Gemini spacecraft be bolted to the rear of a spare TDA to provide a 'passive' docking target for Gemini 8. George Mueller was enthusiastic because by using off-the-shelf hardware it would be cheap to build. On 5 December, Charles Mathews sent the plan to Robert Seamans, who promptly authorised the fabrication of this 'Augmented Target Docking Adapter' (ATDA). Flight plans for Gemini 8 were drawn up by Houston's Flight Crew Operations Division to employ either target. The ATDA was delivered to the Cape in early February 1966, and put in storage. GATV-5003 had been delivered a few weeks earlier, but the recertification of its engine in a test rig was proving frustrating. On 12 February, after six successful firings, the engine suffered a 'hard' start, but the cause was traced to alcohol contamination. Another engine was in place by 1 March, and four days of intensive trials served to certify it flight-worthy. A few days later, this Agena was mated with its Atlas, which was already on Pad 14. The Cape was working hard to integrate all the various strands of activity required to mount the dual launch on the planned date of 16 March 1966.

An ambitious EVA plan

In late 1964, the plan had been for the first Gemini EVA to be just a demonstration of opening and closing the hatch, but following Alexei Leonov's pioneering spacewalk it was decided to have Ed White match this. Whilst Gemini 5 was to attempt an eight-day mission, there was no EVA assigned, and even if one had been planned the fuel cell problem would have prompted its cancellation. There was an early suggestion that Tom Stafford should exit Gemini 6 to retrieve an experiment package from the first Agena docking target, but Wally Schirra dismissed spacewalking as a distraction from the primary objective of achieving the rendezvous and docking. When the Agena blew up and Gemini 6 was recast as a rendezvous with Gemini 7, a proposal based on a study by Bill Schneider, the Gemini Mission Director, called for Jim Lovell to swap places in space with Stafford, but Frank Borman rejected this because he did not wish to give up the 'soft' suits that they were to wear on their marathon mission, which were not designed for extravehicular activity. As a result, therefore, NASA's second EVA fell to Dave Scott. Once Gemini 8 had undocked from its Agena target vehicle and was station-keeping alongside it, he was to exit. Like White, his objective was to assess mobility and stability using a Hand-Held Manoeuvring Unit (HHMU), but in this case using a backpack housing a tank containing fifteen times as much gas as had been available to White in order to make a comprehensive test, and whereas White's 'gun' had used oxygen, Scott's would use freon, which had a higher specific impulse. As it would be impractical to carry this bulky EVA Support Package (ESP) in the cabin, it was to be stowed at the rear of the adapter module, and Scott would have to retrieve and don it, a preliminary that was not expected to pose a problem because White's experience had been very positive.

To enable Scott to operate independently of his spacecraft, an EVA Life Support System (ELSS) had been developed. Whilst the chestpack that White had worn had incorporated a 10-minute emergency-oxygen tank, its primary function had been to ventilate his suit using oxygen drawn from the spacecraft's Environmental Control System via the umbilical. The new chestpack, which was considerably bulkier, was designed to enable Scott to switch over to a tank of oxygen in the ESP in order to enable him to disconnect from the spacecraft. He was to don the ELSS and a 25-foot umbilical prior to opening the hatch, go to the rear of the adapter and strap himself into the ESP. After switching to the backpack's oxygen supply, he would disconnect from this umbilical and link it to a 75-foot tether retrieved from inside the adapter to create, in effect, a 100-foot line. These tasks would be tricky because the field of view through the suit's visor would be limited, and the lack of tactile sensitivity through the EVA gloves would make working 'by feel' difficult. If Scott got into trouble, he would have to switch to the emergency oxygen tank in the ELSS and beat a hasty retreat to the safety of the spacecraft's cabin. The risks involved were emphasised by an incident in training. When Neil Armstrong attended a briefing on the ELSS by McDonnell the start of September 1965, he called for testing it in the altitude chamber in order to confirm that the chestpack and the backpack would integrate properly. This test revealed that a valve in the ELSS could freeze. In fact, this was a 'single point failure' because it would not only inhibit the flow of oxygen to the suit irrespective of whether it was drawing from the spacecraft or from the ESP, but also from its own emergency supply. As a result, a heater was installed to prevent the valve from freezing. It was an object lesson in the value of realistic environmental testing.

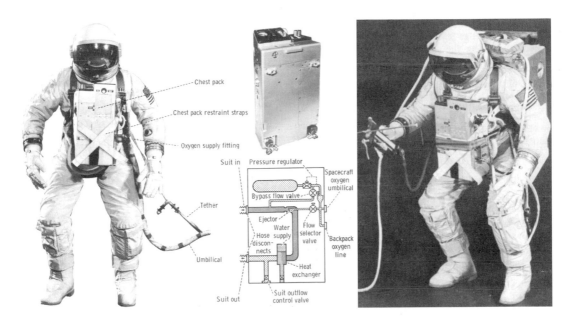

The EVA Life Support System assigned to Dave Scott was much more sophisticated than the chestpack used by Ed White, and the EVA Support Package backpack that he was to retrieve from the rear of the spacecraft's adapter contained a tank of freon to permit a more comprehensive assessment of the 'zip gun' Hand-Held Manoeuvring Unit than was feasible during White's brief spacewalk.

If Scott managed to don the ESP, he was to start by verifying that he could move about purposefully using the HHMU, and then attempt some 'work'. Three of the 10 experiments assigned to this mission were to be undertaken outside. One experiment was to measure the radiation level outside the spacecraft, particularly while passing through the 'South Atlantic Anomaly', a zone off the coast of Brazil where the inner van Allen radiation belt dips down towards the ionosphere, in which astronauts might be expected to receive a higher radiation dosage than at other times in their orbit. Scott was also to 'fly' over to the Agena in order to retrieve and replace a micrometeoroid package· This would be a welcome demonstration of the utility of spacewalkers. NASA had Pegasus satellites with large 'wings' of impact detectors to report micrometeoroid statistics by radio telemetry, but retrieving this package would enable the impacts to be studied directly. As yet, no object had been retrieved and returned to Earth. With a little luck, a future mission would retrieve the package that Scott installed as a replacement.

As this artist's impression shows, one of Dave Scott's EVA tasks was to test a 'reactionless' power tool.

Finally, there was a 'work station' on the side of the adapter where Scott was to assess a torqueless wrench that was designed for use in weightlessness. Scott was to compare tightening and loosening bolts while floating freely and while wearing a variety of tethers for stabilisation. As this experiment was for the Air Force, NASA wanted Scott to repeat the tasks utilising a standard hand-wrench to provide additional 'data points'.

Whereas White had been ordered inside prior to orbital sunset, the fact that this ambitious plan was expected to take 2 hours meant that Scott would have to remain out while the spacecraft passed through the Earth's shadow, so it promised to be the highlight of the mission. As their mission patch, they chose a prism to highlight the 'full spectrum' of their assignments.

The patch for Gemini 8 was intended to represent the 'full spectrum' of tasks: rendezvous, docking and spacewalking.

Go for docking!

The Atlas carrying GATV-5003 lifted off on time at 10:00 Cape time on Wednesday, 16 March. Some 650 nautical miles down the Eastern Test Range, at an altitude of 120 nautical miles, the sustainer engine shut down and the vernier rockets refined its trajectory, then the payload was released. The Agena coasted for 30 seconds and then fired its 16-pound-thrust 'cold gas' SPS thrusters for ullage to settle the propellants prior to starting the PPS. It was at this point that Gemini 6 had lost its rendezvous target. In Houston, Mel Brooks, monitoring the telemetry at the Agena console, reported that the PPS had successfully ignited, and the Spacecraft Test Conductor in the Pad 19 blockhouse told the astronauts that it was burning smoothly.

"Very good," Armstrong acknowledged.

Pyrotechnics fired to split the aerodynamic shroud lengthwise, and the two sections fell away to expose the TDA on the nose of the Agena. The engine underperformed slightly, so the autopilot burned it for 24 seconds longer than nominal in order to establish the desired speed, and when it shut off it was within 3 feet per second of that required for the planned circular orbit at an altitude of 161 nautical miles.

In Houston, John Hodge lit up a cigar to celebrate. He had recently taken over as chief Flight Director from Chris Kraft, who was now managing the preparations for Apollo.

"It looks like we have a live one up there for you," the Spacecraft Test Conductor told Gemini 8.

"Good show!" Armstrong replied. On being informed that telemetry indicated that the Agena had deployed its various antennas, he added enthusiastically, "Beautiful, we'll take that one."

"That's just what the doctor ordered!" Scott agreed.

Neil Armstrong leads Dave Scott up the ramp on Pad 19 on 16 March 1966.

The radars of the World-Wide Tracking Network monitored the Agena through its first revolution to determine its orbital parameters, to enable the best time to initiate the chase to be calculated. Gemini 8's count was synchronised during a 'hold' at T–3 minutes, and then launched without incident at 11:41.

"You're 'Go' for staging," advised Jim Lovell, the Houston CapCom. The commander of the previous mission usually served as CapCom for this phase, but in early February NASA had dispatched Frank Borman and Wally Schirra on a 'goodwill' tour of the world, and they were now on their final leg, crossing the Pacific.

"Ignition! Staging looks good," Armstrong reported. "We're having a real fireball, here," he added to Scott on the intercom when a brilliant flash lit up the windows.

"Yes," Scott agreed.

"You have Guidance Initiate," Lovell advised when the second stage began to refine the trajectory. Its progress was monitored by the Flight Dynamics Officer, Cliff Charlesworth, who would soon be promoted to Flight Director's status.

"The second stage is a real good machine," Armstrong enthused to Lovell. To Scott, he added, "I was going to say 'Cadillac', but I guess I'd better not say that."

"How about that view," Scott said appreciatively.

"Fantastic," Armstrong agreed. "They were right." They were both space 'rookies', but had been thoroughly briefed by their predecessors.

As they attained 85 nautical miles altitude, the Titan accelerated through 80 per cent of the speed required to achieve the desired orbit. Even if the Titan failed now, they would be able to separate and use their own OAMS thrusters to limp into a low orbit. However, the Martin Company's vehicle shut down right on time.

"We've had SECO," Armstrong reported.

"Take your time, buddy," Scott said, as Armstrong counted down the 20 seconds before firing the OAMS to move clear of the spent stage, which this time was ignored as junk.

The radars of the Eastern Test Range had determined that the plane of Gemini 8's orbit was almost exactly coincident with the Agena's, and Canary Island reported that the 86 by 147 nautical mile initial orbit was within 1 mile of that intended. They were well positioned to chase the Agena, which was just over 1,000 nautical miles ahead.

"You're 'Go' for a nominal 'm=4'," Lovell advised via Ascension Island. It was to be a straight re-run of how Gemini 6 caught Gemini 7 on the fourth revolution. "There will be a slight plane change, but we'll give you that information later."

"All systems look okay," Armstrong reported, having just completed the post-insertion checklist to verify the spacecraft's systems.

As Gemini 8 flew towards Australia, Hodge called Carnarvon and asked, "Do you have any questions?"

"Negative," the CapCom assured. He had listened to the launch commentary over the loop. When the Agena rose above his horizon for the second time he checked its telemetry and told Houston, "The Agena looks real good." Then Gemini 8 made its first appearance. "How're y'all doing?"

"Just fine," Armstrong replied.

"We're showing you real good." He added that they were 'Go' for '16-1', meaning that they were cleared to fly a 24-hour mission at least. "You like it up there, right?"

"You bet!" Scott replied.

"It's alright," Armstrong allowed.

"Very nice," the CapCom replied wishfully.

"You couldn't do any more for us than you have," Scott thanked the CapCom after he had relayed some updates. "We've got everything going for us now."

"That's what we're here for, isn't it."

As Gemini 8 approached Hawaii, Arda Roy, the CapCom, called Houston. "Be advised that Wally Schirra – callsign 'Gemini 76' – is attempting to contact Gemini 8. He's inbound to Honolulu."

"Howsabout that!" Hodge replied.

"Gemini 8, Gemini 76," Schirra called, "do you read?" As there was no reply Roy called the spacecraft to announce his acquisition of signal.

"We have you in sight down there," Scott replied. "It looks like a nice day."

"It's beautiful weather here," Roy confirmed. After giving the data for the first burn, he alerted the astronauts to Schirra's attempts to establish contact.

"We'll be standing by," Scott assured.

"Hiya, Dave?" Schirra called, but he could not make contact before the spacecraft flew out of range.

Although communications via Guaymas in Mexico were noisy, Lovell gave the 'Go' to perform the first manoeuvre at perigee over the Gulf of Mexico, just south of New Orleans. As with Gemini 6, the manoeuvre sequence was being calculated by the ground on the basis of tracking by the WWTN, and the initial solution was refined on an ongoing basis. As the insertion orbit was a little over a mile higher than planned, this burn was to be retrograde to trim the speed by 3 feet per second and slightly lower the apogee. As they flew across the continent, they aligned Gemini 8 'small end forward' (SEF) so the forward-firing thrusters would slow them down. "Keep an eye out the window," Armstrong urged, "and see what it looks like with this firing." His attention would be on the instruments for the 5-second burn and he wanted Scott to observe the plumes from the thrusters. As they headed out over the Atlantic for the second time, they had the first meal. The dehydrated powder in the plastic bag did not reconstitute very well using the 'gassy' water that emerged from the dispensing nozzle. The second manoeuvre was made over the Indian Ocean. Lasting over a minute, this prograde burn lifted the perigee by 28 miles in order to set up the phasing of the chase. "We completed the burn on time," Armstrong reported on establishing contact with Carnarvon.

"You're looking real fine," the CapCom advised. "And your Agena is looking real good, too." Their target vehicle was now 375 nautical miles ahead.

"Thank you," Scott acknowledged.

"We'll hang loose here," the CapCom said, "and keep quiet." Unless he had something specific to pass up, he would remain silent. Of course, if the astronauts wanted to chat, he would oblige them.

Radar tracking had established that Gemini 8's orbital plane was inclined at an angle of 0.05 degrees to that of the Agena. As had Schirra, Armstrong corrected this several minutes before reaching Hawaii.

"Ask him how the burn went," Hodge ordered Hawaii.

"The burn was on time, and the residuals nulled," Armstrong reported.

As they approached the coast of Mexico, Lovell called through Guaymas with an urgent correction. "We want to give you another burn here, very shortly. Standby to copy." Taken by surprise, the astronauts had to scramble to retrieve the flight plan to write the burn data. Lovell's "very shortly" turned out to be no understatement, because the manoeuvre had to occur 90 seconds after his 'heads up'. As Lovell called the final few seconds in countdown, Armstrong hastily aligned the spacecraft SEF. The increment was just 2 feet per second, but refining the apogee by 1 nautical mile prior to the co-elliptic manoeuvre would significantly improved the rendezvous 'solution'.

"We got it!" Armstrong confirmed.

"Sorry for that," Lovell apologised afterwards, "but we had a malfunction here, and we had to get you the burn in a hurry." A minute later, he was back: "And be informed that the Agena is now configured TDA-north." The Agena usually travelled with its TDA facing the direction of travel, but it had been yawed through 90 degrees for Gemini 8's approach so as to present the best view of its blinking acquisition lights. Just before the Eastern Test Range lost them, Lovell read up the data for the co-elliptic burn.

After configuring the spacecraft, Pete Conrad and Dick Gordon, the backup crew, had watched the launch from in the blockhouse and then flown home. They now joined Lovell at the CapCom's console. Gordon was keen to monitor the terminal phases of the rendezvous. For this flight, Hodge and Gene Kranz were to alternate as Flight Director on 12-hour shifts. At this point, Kranz appeared, and got a feel for how the mission was progressing by reading Hodge's console log.

The *Rose Knot Victor*, stationed in the South Atlantic, commanded the Agena to switch on its acquisition lights.

"We're getting intermittent lock-on," Armstrong reported. The radar transponder on the Agena was amplifying and returning the signal broadcast by the Gemini spacecraft. As yet, however, the reply was not strong enough for the computer to reliably process its range and range-rate data. As they closed within 180 miles, the signal firmed up. Armstrong raised his spacecraft's nose towards the target in order to boresight his optical reticle with the radar to see what he could see.

Paul Haney, the Public Affairs Officer in Houston, emphasised the radar's importance to the rendezvous. "The pilots say that if they had to lose any of the several things involved in a radar mission – that is, the platform, the computer or the radar – the one they would rather keep over everything is the radar."

The co-elliptic burn made while passing over the Tananarive relay site on the island of Madagascar nudged Gemini 8 into an orbit a constant 15 nautical miles beneath that of the Agena. Being lower, it would gradually catch up with the target vehicle in terms of central angle with the timing designed to establish the conditions for the terminal phase.

"We've got a visual on the Agena at 76 miles," Scott reported as they crossed Texas for the third time. He began taking sightings of the target. "At least," he mused, "we have *some* object in sight, something that looks like it would be the Agena."

"The Agena or Sirius!" Lovell teased, harking back to Schirra's uncertainty on sighting Gemini 7.

"It could be a planet," Scott admitted, but as they tracked the object it became evident that it was indeed the Agena reflecting the Sun.

After the WWTN radars had measured Gemini 8's circularised orbit, Lovell read up the refined data for the TPI burn. Armstrong was keeping the spacecraft's nose facing its target. By maintaining the Agena in the centre of his reticle as the range reduced, Gemini 8's nose was pitching progressively upwards – the terminal phase was to be initiated when the angle reached 27.4 degrees.

"Houston, Gemini 8," Armstrong called. There was no answer, because they had passed beyond Texas's range a minute previously. The Earth was turning on its axis, so the ground track migrated 22.5 degrees west with each spacecraft revolution, so by now their northerly arc was barely clipping the southern United States.

"Boy," Scott observed, "we didn't stay with them long, did we."

Gemini 8 flew into the Earth's shadow while crossing the South Atlantic. Armstrong was still able to hold the Agena centred in the reticle by its acquisition lights but, as he informed the *Rose Knot Victor*, the target was "very hard to see." The TPI burn was performed when the angle was just right – a minute and a half later than predicted – to start the gentle climb.

"You needn't answer," Lovell called via Tananarive. "We're standing by."

"We burned TPI on 'closed loop'," Armstrong reported.

And that was it. Lovell did not want to disturb them in this critical phase. Dick Gordon was taking note of Scott's running commentary of sightings, and plotting the approach on a chart. Armstrong continued to pitch up in order to hold the Agena centred in his reticle and made small burns to prevent it from drifting against the background stars, thereby ensuring that they flew the initial 26 miles of the approach along the line-of-sight at the time of TPI.

"We're on the way!" Scott reported when the *Coastal Sentry Quebec* announced that it was listening out.

Moments later, with the Agena only 2 miles away and the range-rate down to 50 feet per second, they flew into daylight. All of the earlier manoeuvres had been designed to establish this situation. The terminal phase had to be flown in darkness so that Armstrong could fix the Agena against the stars during the line-of-sight approach, but the braking had to be done in daylight so as to minimise the risk of a collision. Scott used his three-digit 'encoder' to order the Agena to switch off its acquisition lights. The fact that it did so confirmed that the radio command link was working. The illumination of the Message Acceptance Pulse (MAP) light on his console provided further confirmation.

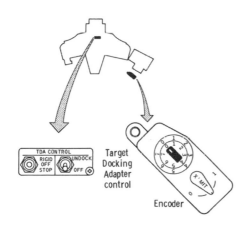

The spacecraft assigned to docking missions had additional controls, including the 'encoder', which sent three-digit instructions to the Agena. The first two digits were set by rotating the two dials and the act of selecting either '0' or '1' issued the command.

"Okay, I'm going to start braking a little," Armstrong decided. He intended to brake in a series of steps, starting at 1.8 miles – about four times further out than Schirra had started to brake. A minute and a half later, the range was 8,000 feet and the range-rate was 40 feet per second. "Let's back off a little bit more," Armstrong said, as they passed through 6,000 feet.

"31 feet per second," Scott reported, "a little high."

"I'll brake down a little bit more."

After several more minutes of braking brought them within 1,000 feet, Scott activated a 16-mm movie camera that he had mounted in his window, to document their approach. The Agena was in precisely the programmed attitude.

"That's just unbelievable," Armstrong

mused.

"Outstanding job, coach."

"Way to go, podner."

"You did it, boy," Scott insisted. "You did a good job."

"It takes two to tango," Armstrong countered.

"Look at that sucker!" Scott remarked as they coasted in. The Agena was rock steady.

"See the dipole?" The Agena had several antennas. The dipole projected 'up' from just behind the TDA's collar.

"Do I ever!" Scott insisted. He had grabbed a Hasselblad, and was snapping pictures of the Agena. The clarity of the detail was stark. "I can I see everything on that fellow."

"Standing by for any rendezvous remarks," Lovell called up via the *Range Tracker*, a ship stationed 1,500 miles west of Hawaii.

"You tell 'em," Scott said to Armstrong.

Armstrong keyed the radio, "We're station-keeping on the Agena, at about 150 feet."

Unfortunately, the radio link was so bad that Lovell could not read the transmission, so he advised that they would wait until Hawaii.

"How's it look?" Hodge asked Hawaii a minute later.

"Both vehicles are 'Go', Flight."

"Ask him how he's doing!"

"How're you doing?" Arda Roy relayed.

After rendezvousing with the Agena, Gemini 8 manoeuvred around to line up with the axis of its docking adapter.

"Station-keeping at 150 feet," Armstrong repeated.

"Roger," Roy acknowledged matter-of-factly.

"Our Agena," Scott began, for Houston's benefit, "is in fine shape. The TDA is out. The whiskers are sticking out, as expected, on the TDA." The wire 'whiskers' were to discharge any static electricity on contact that had accumulated as a result of travelling through the ionosphere. "The dipole is up. The engine looks good. And we have turned the acquisition lights off."

With a PQI of 55 per cent, the rate of fuel consumption had closely matched the prediction based on Gemini 6's rendezvous with Gemini 7.

"It flies easy!" Armstrong enthused. "I'd love to let you do it, but ..."

"Oh, no!" Scott retorted.

"I think I'd better get my practice in while I can."

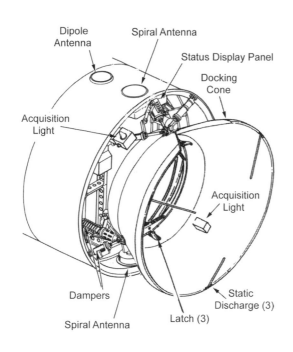

Dipole Antenna
Spiral Antenna
Status Display Panel
Docking Cone
Acquisition Light
Acquisition Light
Dampers
Static Discharge (3)
Spiral Antenna
Latch (3)

The conical collar of the Agena's Target Docking Adapter was extended to accept the nose of the Gemini spacecraft, and once the three capture latches had engaged the collar was drawn in and rigidised to lock the two vehicles together.

"I'll have my chance!" Scott assured. The flight plan allowed him to fly the spacecraft in proximity operations later on, during redocking tests. "I wouldn't even take it if you gave it to me."

"This station-keeping," Armstrong mused, "there's nothing to it." He was slowly circling the Agena at 80 feet, manoeuvring to enable Scott to inspect the Status Display Panel on the TDA using his telescopic sextant.

The Public Affairs Officer was impressed by the nonchalance. "All in all," he said, "the pilots are acting extremely 'ho-hum' about the whole thing."

As they flew on, Armstrong lined up with the TDA, closed in and halted. "We're sitting about 2 feet out," he told the *Rose Knot Victor* CapCom, Keith Kundel, once they had risen above the ship's horizon.

"Is he docked?" Flight prompted.

"Negative," Kundel replied, "he's not docked yet." Then he called Gemini 8. "You're looking good. Go ahead and dock."

"Let me know when he does," Hodge ordered.

"We're going to go ahead and dock," Armstrong called. He pulsed the aft-firing thrusters to ease Gemini 8's nose into the conical TDA at a speed of 6 inches per second. Contrary to expectation, there was no electrical discharge. To accommodate an off-axis entry, the collar was free to pitch and yaw within its mounting. After seven hydraulic dampers had absorbed the energy of the contact, three mooring latches engaged and an electric motor activated the gear boxes that 'rigidised' the mechanism and locked the Gemini on the axis. When the light on the SDP illuminated to show that the mechanism was rigid, Scott commanded the motor off. Because the Agena's TDA was facing north, the Gemini was facing south. The docking occurred as they crossed the evening terminator, so whilst Scott viewed the side that caught the last rays of the Sun, from Armstrong's viewpoint the Agena was mostly shadowed and it had been almost a night-time docking.

Dave Scott took this picture moments before Neil Armstrong drove their spacecraft's nose into the TDA to achieve the first-ever docking in space.

"Flight, we're docked," Armstrong announced matter-of-factly. "It was a real smoothie."

In Houston, the flight controllers raised their thumbs and cheered. The primary objective of the Program had been accomplished.

"Congratulations!" Kundel praised.

"You couldn't have the thrill down there," Scott pointed out, "that we have up here."

"For your information," Armstrong resumed his report, "the Agena was very stable, and at the present time we're having no noticeable oscillations at all."

"Very good," Kundel acknowledged.

Rendezvous (m=4)	Rev	Gemini 6	Gemini 8
1. Height adjust	1	01:34:02	01:34:37
2. Phasing	2	02:18:00	02:18:25
3. Plane change at node	2	02:42:07	02:25:00
4. Height adjust	2	03:03:19	03:03:41
5. Co-elliptic	3	03:47:37	03:47:35
6. TPI	4	05:17:54	05:13:52
7. TPF	4	05:48:40	05:45:37

A wild ride!

After inspecting the Agena's telemetry, Kundel on the *Rose Knot Victor* told Gemini 8 that it would be necessary to reload the Stored Program Controller because Hawaii had not been able to verify its upload. Armstrong was reluctant for the Agena to be reloaded while they were in docked configuration, but relented when promised that the update would be triple-checked prior to being transmitted. Meanwhile, Scott used the encoder to order the Attitude Control System (ACS) at the rear of the Agena to fire its cold-gas jets to yaw the docked combination through 90 degrees as the first in a series of manoeuvres designed to assess the strength of the TDA. The first yaw around was at the sedate rate of 1.5 degrees per second, taking just over a minute. Because this rotated the Agena from TDA-north to TDA-forward, Gemini 8 was now flying backwards. Although Armstrong was impressed by the crispness with which the Agena initiated and ended the turn, he was disappointed that it had ended up slightly pitched down.

When Lovell called via Tananarive, he warned that since the *Rose Knot Victor* had not been able to verify the SPC upload, they should proceed with caution. "We believe we have the load in, and we would like you to enable the SPC and let the Agena start. But if you run into trouble, and the ACS of the Agena goes wild, just turn it off and take control with the [Gemini] spacecraft."

"We understand," Scott acknowledged as the Tananarive relay faded.

As they flew on in darkness, Armstrong read a string of three-digit codes from the flight plan, and Scott used the encoder to transmit them to the Agena to prepare it for the second manoeuvre. On looking up afterwards, Scott saw that the '8-ball' attitude indicator showed that they had rolled 30 degrees to the left. Neither man had felt this excursion, and because they had the cabin lamps full on they could not see anything outside and there had been no visual cues. Suspecting an Agena malfunction, Scott commanded off its ACS, deactivated its horizon sensor, and inhibited the function that maintained its orientation fixed relative to the Earth. Then Armstrong reactivated the OAMS, and within a minute restabilised the docked combination. When it promptly began to drift again, he asked for the ACS to be re-engaged to damp out the spurious motion, but this proved ineffective.

By this point, some 25 minutes after they had docked, they were approaching the dawn terminator. On spotting a 25-foot-long persistent plume projecting from one of the Agena's pitch thrusters in the growing light, Scott realised that the Agena was rapidly expending its propellant and he shut off its ACS again. Believing the Agena to be at fault, they decided to undock. As soon as Armstrong had slowed the rates, Scott commanded the TDA to release them and Armstrong jetted the forward-firing thrusters for 5 seconds for a rapid withdrawal. A 16-mm movie camera in Armstrong's window was still operating, and caught this action. Remarkably, the TDA had withstood the stresses of the spin up and the attempts to cancel it without suffering oscillations, and so had passed a much more severe 'bending moment' test than planned.

Within seconds of moving clear, Gemini 8 spun up at an even faster rate than before and Armstrong realised that the fault was in his own spacecraft. He suspected that a short in the hand controller was commanding a thruster to fire continuously. The accelerating excursion was mostly roll, but there was also an element of yaw. As he struggled to regain control, the Propellant Quantity Indicator rapidly fell to 30 per cent. They were in serious trouble!

"How does it look?" Hodge enquired casually when he decided that the *Coastal Sentry Quebec* should have acquired the signal.

"We're indicating spacecraft-free!" Jim Fucci replied in astonishment. "I'm going to call the crew now."

"Say again?" Hodge was sure that he must have misheard the report of 'spacecraft-free', which would indicate that Gemini 8 had withdrawn from the Agena.

"They're not docked!" Fucci confirmed.

"Call them!" Hodge urged.

"We have a serious problem," Scott pointed out, when the ship announced its presence. "We've disengaged from the Agena."

"What seems to be the problem?" Fucci asked.

"We're rolling up," Armstrong began, "and we can't turn ...", at which point his voice faded, picking up a moment later with, "... a continuously increasing left roll." The link was intermittent because the antennas on the spinning vehicle could not maintain contact with the ground. And Fucci was getting little hard information from the telemetry stream.

"Did he say he couldn't turn the Agena off?" Hodge asked Fucci.

"No, he said he had separated from the Agena," Fucci corrected. "He's in a roll and he can't stop it."

It was clear that whatever problem Gemini 8 had encountered, the crew would have to overcome it themselves. With the spacecraft spinning at a rate of one cycle per second and accelerating, they had severe tunnel vision, and were near to blacking out. Recognising that they were fast running out of time, Armstrong decided on drastic action. He cut the power to the OAMS control system and pulled the circuit breakers for all of the thrusters, and then he armed the Re-entry Control System (RCS) mounted in the spacecraft's nose. This contained two 'rings' of eight jets, each of 25-pound-thrust. Although intended to stabilise the vehicle for re-entry, the RCS was now the only means of regaining control. However, whereas the propellant in one ring was deemed sufficient for re-entry and the second provided a healthy degree of redundancy, there was a risk that stabilising the rapidly spinning spacecraft would consume so much propellant that there would be insufficient left for re-entry.

"We're in a violent left roll here," Scott repeated for anyone who was listening, and then announced the remedy: "At the present time, the RCS is armed. We apparently have a stuck hand controller."

"They seem to have a stuck thruster," Fucci told Houston.

"Did I hear him say he may have a stuck hand controller?" Hodge asked.

"That's affirmative."

After another minute and a half of silence, Scott reported the good news. "Okay, we're regaining control of the spacecraft slowly."

"We're showing pretty violent oscillations in roll," Fucci told Hodge when he started to receive telemetry.

"We're pulsing the RCS pretty slowly," Armstrong called. "We're trying to kill our roll rate."

"The Agena is tumbling violently also," Fucci reported, after checking its telemetry.

"I understand you have ACS 'Off', right?" Hodge wanted to verify his understanding of the situation.

"That's affirmative."

"You'd better get the ACS 'On', and get it in FC-1." Hodge wanted the *Coastal Sentry Quebec* to reactivate the Attitude Control System and instruct the Agena to stabilise itself. Despite the pace with which the crisis had developed, Scott had had the presence of mind to disengage his encoder, thereby ensuring that the Agena would be able to accept commands from the ground.

"Find out how much RCS fuel he's used," Hodge called, "and if he's just on one ring."

"We're on one ring," Armstrong replied when Fucci relayed the call, "trying to save the other ring."

"How're you doing?"

"We're working on it!" Armstrong assured. "The docked combination just took off in yaw and roll," he added by way of an explanation.

"See if you can find out their relative positions," Hodge urged Fucci. He was concerned that the two vehicles might collide.

"Can you see the Agena now?" Fucci asked. There was no reply.

Armstrong continued his account. "We turned ACS 'Off', then we turned on OAMS and tried to stabilise – in so doing, we may have burned out our roll-left thruster?"

"Do you have visual contact with the Agena right now?" Fucci persisted.

"No," Scott replied. "We haven't seen the Agena since we undocked." He then pointed out that prior to undocking he had ordered the Agena's ACS 'Off' and had got a MAP light, indicating that it had done so.

"Flight," Fucci called Hodge, "he said that the spacecraft and Agena combination both started to yaw and roll violently, so they initiated ACS 'Off' and undocked from the thing, then they lost sight of it."

"Did you get ACS *back on*?" Hodge asked.

"Affirmative."

Having stabilised his spacecraft using the RCS, Armstrong then powered up the OAMS and pushed in the circuit breakers one by one to identify the faulty thruster – it was one of the yaw thrusters on Scott's side. The multiple-axis excursion derived from the fact that the yaw thrusters were mounted in pairs, and when one fired without the other it induced a roll component into the yaw. The valves of the thrusters were activated by electrically powered solenoids which were held at high potential and 'grounded' when selected. The OAMS had been serviced three days prior to launch, and it was surmised that one of the wires had come loose and was causing the thruster to fire erratically when not selected, and not to fire when commanded. The evolution of the attitude excursion from a combined yaw and roll to nearly pure roll was because the yaw component was cancelled out as the vehicle rolled, giving rise to a progressively worsening 'coning' motion. Because the 'ground to fire' logic had allowed the thruster to 'fail on' whilst 'logically off', Armstrong's only option in the very short time available was to power off the entire OAMS system, which in turn had obliged activating the RCS to stabilise the spacecraft.[1]

[1] The OAMS system was subsequently revised for apply-power-to-fire.

Emergency return

As Gemini 8 passed beyond the *Coastal Sentry Quebec's* range, the mission transformed from the celebration of the historic docking to an emergency, and because the RCS had been activated the mission rules called for a return to Earth as soon as possible. After a discussion with Kranz, Hodge invoked the rule and because Kranz was to have handled the re-entry at the conclusion of the mission, Hodge, who had been on duty for 11 hours, relinquished control.

Arda Roy in Hawaii passed on the decision. "Be advised, we're planning to come into a dash-3 area." The 'dash-3' landing areas were for emergencies only, with minimal recovery forces on station. "We're looking at rev 6 or 7."

Armstrong reported that the OAMS was down to 22 per cent, and most of the RCS ring that he had activated had been consumed.

"Do you have anything further, Flight?" Roy asked as the spacecraft dropped towards his horizon.

"Ask if they have any idea where the Agena is," Kranz said.

"We saw it about 10 minutes ago," Armstrong replied, referring to the Agena, once Roy had relayed the query. "It looked to be a mile or so underneath us."

"Could you give us the sequence of events?" Roy asked at Kranz's prompting, but the signal faded out soon after Armstrong started to respond.

In Houston, Paul Haney, the Public Affairs Officer, announced the result of the Flight Directors' conference. It had been decided to use the 7-3 area, which was some 500 nautical miles east of the island of Okinawa and 630 miles south of Tokyo in Japan – in fact, not far from the *Coastal Sentry Quebec's* station. Whilst the destroyer *USS Leonard F Mason* had been ordered to the aim point, it would have to steam 160 miles, and so would not arrive on scene until several hours after Gemini 8 splashed down. To provide immediate assistance, a pair of HC-54 Rescuemasters aircraft flew in, one from the Japanese mainland and the other from Okinawa, to deliver pararescue divers and a flotation collar for the capsule. In addition, an amphibian aircraft was dispatched just in case it became necessary to make an emergency pick up. The good news was that the weather was excellent and the sea state was calm. The *Mason* was to deliver the astronauts to Naha on Okinawa, from where they would be flown home.

Meanwhile, Kranz told the *Rose Knot Victor* to note the time that it acquired the signal from each spacecraft in order to measure the separation, and the 2 degree angular difference put them 5 miles apart. Once he raised Gemini 8, Kundel relayed the news that it had been decided to make a 7-3 recovery and that the de-orbit burn would be in 75 minutes, so they would have to hustle to restow everything that they had unpacked and then run through the pre-retro checklist.

"Is your pre-retro checklist complete?" Fucci enquired 40 minutes later when Gemini 8 reappeared over the *Coastal Sentry Quebec's* horizon.

"It is," Scott replied, "we're just finishing up restowage."

Once Scott had retrieved the flight plan, Fucci read the timings for the de-orbit burn and the re-entry sequence. As they made their final revolution, Arda Roy in Hawaii updated the re-entry data and gave a weather report for the recovery area. Armstrong and Scott checked each other's actions to ensure that they made no mistakes. Over the *Rose Knot Victor* in the South Atlantic, Armstrong reported that

Swimmers dropped by aircraft attached a flotation collar to Gemini 8 after its splashdown in a contingency recovery site in the Pacific (top). Neil Armstrong and Dave Scott onboard the *USS Mason* with their Air Force pararescuers: A/2C Glenn M Moore (standing) and A/1C Eldridge M Neal (left) and S/Sgt Larry D Huyett (right).

they had completed the checklist. In Houston, Paul Haney told the anxious journalists that the crew sounded "quite relaxed". Jim Lovell tried in vain to make contact via Ascension Island, and then provided a countdown for the burn 'in the blind' via Kano in Nigeria. The de-orbit burn was made over the Congo in darkness, and Armstrong's confirmation that all four retros had fired prompted a cheer in Mission Control. Barely 10 hours after launch, Gemini 8 was on its way home. It entered the atmosphere over China, still in darkness. Armstrong managed to use Sirius to verify their orientation, before switching his attention to 'flying the needles' in an effort to land at the recovery point. Scott sought external cues. "Do you see water out there?" Armstrong enquired as they neared the dawn terminator.

"All I see is haze," Scott replied. A moment later, he changed his mind. "Oh, yes, there's water!" At least they would not set down on land.

As soon as they splashed down, the HC-54 Rescuemasters began to circle. One dropped swimmers into the 15-foot swell to affix a flotation collar. The astronauts threw open their hatches and chatted with their rescuers, and after the *Mason* eventually drew alongside they eagerly scrambled up a rope ladder onto the deck and watched as the ship's crane lifted the spacecraft aboard.

When Hodge announced his decision to have Gemini 8 make an emergency return to the western Pacific, Deke Slayton had telephoned Wally Schirra, who had just landed in Hawaii on the final leg of his world tour with Frank Borman, and asked him to fly to Okinawa with instructions to "keep them under control", by which he meant keep Armstrong and Scott in isolation in order to preserve their impressions of the mission for the preliminary debriefing, which would be conducted in Hawaii. As soon as the *Mason* tied up, Schirra whisked them off to a helicopter for transfer to the local airport, where an aircraft was waiting to fly them to Hawaii for recuperation in the Tripler Army Hospital.

On arrival at the Cape two days later, Armstrong wryly told the waiting reporters, "We had a magnificent flight – for the first 7 hours."

"All of us are greatly relieved," said President Johnson. Even so, the TV networks had received thousands of telephone calls complaining that the live coverage of the arrival in Okinawa had interrupted re-runs of '*Batman*' and '*The Virginian*'.

Armstrong and Scott received NASA's Exceptional Service Medal "for the remarkable job" that they did, particularly for having "the presence of mind as they were undocking to leave the Agena responsive to ground command and with its tape data intact so that it could be read-out to the ground." Military astronauts received a promotion upon making their first space flight, and Scott became an Air Force Lieutenant Colonel. As Armstrong had retired from the Navy prior to becoming an astronaut, the President ordered that he be awarded an appropriate pay rise.

Chris Kraft pointed out that Armstrong's deactivation of the OAMS and use of the RCS to recover the situation had "proved why we have test pilots in those ships. Had it not been for that good flying, we would probably have lost that crew."

"We learned a lot," noted Robert Gilruth, "and perhaps we learned more than we set out to learn."

In addition to finally achieving a rendezvous and docking, the Program had survived its first in-flight emergency.

In addition to their official portraits, some of the crews had a little fun. Here Dave Scott and Neil Armstrong (in front) pose with their backups Dick Gordon and Pete Conrad. (Sadly, the original photograph was not available.)

Proving the Agena PPS

Over the next two days, Mel Brooks at the Agena console in Houston put GATV-5003 through its paces. Firstly, it fired its PPS to draw its circular orbit out into an ellipse having an apogee of 220 nautical miles and then it circularised this high orbit. When it was fired a third time in order to adjust the orbital plane, the vehicle suffered a yaw excursion and the manoeuvre had the side-effect of further stretching the apogee to 336 miles. When Jerome Hammack, the Agena liaison officer in the Pad 14 blockhouse, said that an offset centre of mass had caused the spacecraft to progressively rotate during thrusting, a revised series of manoeuvres was commanded taking this into account. As the engine itself had performed flawlessly, the Agena was 'parked' in a 220-mile circular orbit in order to serve as an inert 'target of opportunity' in the event that a future mission was permitted to light its Agena's PPS for docked manoeuvres leading to a supplementary rendezvous. In fact, the Gemini 10 crew were already training for such a possibility.

Chapter 7

A Demanding Mission
Gemini 9

Lost crew

In October 1965 Deke Slayton assigned Elliot See and Charles Bassett to the Gemini 9 mission, with Tom Stafford, who was due for launch on Gemini 6, rotating into the backup commander's slot. The postponement of Gemini 6 to December meant that Stafford would not be able to devote his full attention to Gemini 9 until early February 1966, but he did not expect to have any trouble catching up.

The Astronaut Office had a fleet of T-38 Talons to enable its astronauts – every one of whom was a 'fighter jock' who loved fast jets – to maintain their flying proficiency as they crisscrossed the country on visits to contractors. On 28 February 1966, See and Bassett set off from Houston to visit the McDonnell factory in St Louis where Spacecraft 9 was being assembled. The weather at Lambert Field was poor. Upon descending through the 600-foot overcast and finding visibility limited by snow flurries, See opted to 'go around' in order to make a second approach. Following in another T-38, Stafford and Cernan did likewise. See decided to circle beneath the cloud to enable him to maintain visual contact with the runway. Noting that he was losing more altitude than he could afford in the turn, he lit the afterburner to zoom clear. Unfortunately, the aircraft clipped the roof of Building 101 of the McDonnell plant and cart-wheeled into the parking lot beyond, where it was engulfed in a ball of flame in which See and Bassett died. By a cruel twist of fate, they had struck the building in which their spacecraft was being assembled. After making a climbing turn in order to fly around 'on instruments', Stafford and Cernan landed without incident, and without realising that there had been an accident. The first sign that something was wrong came when the control tower asked them to identify themselves – as a means of determining the identity of the lost crew.

On 2 March, when Spacecraft 9 left the McDonnell plant, it was driven by a 'Stars and Stripes' that was being held at half-staff. See and Bassett were buried with full honours in the Arlington National Cemetery in Washington the following day.

Slayton promptly advanced the backup crew to prime – it was the first time that this had been done – and so within three months of returning from Gemini 6 Stafford found himself on the prime crew for a second mission.[1] Faced with the requirement to find a new backup crew, Slayton simply advanced Lovell and Aldrin from this function on Gemini 10.

---

[1] Tom Stafford became the first astronaut to fly two Gemini missions. If it were not for the loss of the original Gemini 9 crew, this distinction would have gone to John Young on Gemini 10.

Elliot See and Charles Bassett were assigned the Gemini 9 mission.

Rendezvous plans

In the summer of 1965, on the presumption that Gemini 6 and Gemini 8 would be able to demonstrate the 'm=4' case of orbital rendezvous, the plan was for the next two missions to attempt the faster 'm=3', because the Apollo planners wanted a lunar module to rendezvous with its mothership on the third revolution after lifting off from the Moon. Detailed planning started in September, a month before the Gemini 9 crew was selected. On 'm=4', the various manoeuvres had been timed to enable the World-Wide Tracking Network radars to determine the effect of a burn and supply input to the calculation of the next burn. In order to compress this timescale, instances of several classes of orbital adjustment were to be combined into single manoeuvres. This was applied to the post-SECO burn. Instead of simply firing the aft thrusters to move clear of the spent Titan stage, Gemini 9 would make an Insertion Velocity Adjustment Routine (IVAR) burn to correct the launch vehicle's in-plane velocity error. A phase adjustment at the first apogee would raise the perigee to establish the general timing for rendezvous. A 'corrective combination' at the start of the third revolution would refine the phasing, tweak the apogee and eliminate any out-of-plane error. The co-elliptic manoeuvre to circularise the orbit just below that of the target would be made 90 degrees later, with the terminal phase initiating the interception later in that revolution.

Early on in planning, consideration was given to having Gemini 9 use the Agena's Primary Propulsion System to rendezvous with the Agena left by Gemini 6, but when this was lost it was decided to leave such docked manoeuvres to a later mission. In any case, there would be no time for such manoeuvres because Gemini 9 inherited several tasks from Gemini 8 which were to have tested procedures for Apollo. It had initially been intended for both the Apollo mothership and lunar module to have rendezvous radars, but in February 1965 the radar on the mothership had been deleted in order to reduce that vehicle's mass. This gave rise to two 'contingency situations' – one in which the lunar module's radar failed and it proceeded to rendezvous by tracking its mothership optically against the dark sky, and the other in which the disabled lunar module managed only to establish an unstable low orbit and the mothership descended to rendezvous by optically tracking it against the backdrop of the lunar surface, either lit or unlit. After disengaging from its Agena, Gemini 9 was to rehearse these tasks by re-establishing a co-elliptic configuration below or above the target vehicle, respectively, and attempt to re-rendezvous by using optical tracking to compute the manoeuvres to initiate and pursue the terminal phase. If this proved to be straightforward, the Apollo planners intended to argue for the deletion of the lunar module's radar in order to lighten that craft.

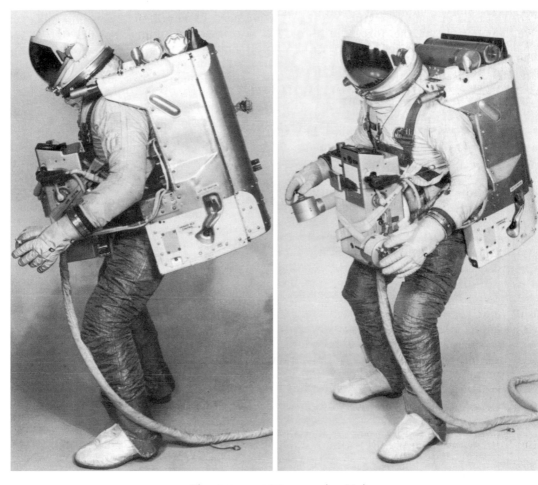

The Astronaut Manoeuvring Unit.

Buck Rogers

The Air Force was hoping to use NASA's Gemini spacecraft for its own 'Blue Gemini' project. This called for astronauts to exit their spacecraft and inspect satellites, particularly Soviet satellites. There were two schools of thought on mobility during EVA: one favoured the 'zip-gun' of the Hand-Held Manoeuvring Unit (HHMU), which was simple to make but could induce rotations; the other favoured a backpack that was capable of imparting 'pure' motions in five degrees of freedom by delivering its impulse through the wearer's centre of mass. In effect, an astronaut equipped with such a Buck Rogers jetpack (or 'Astronaut Manoeuvring Unit' – AMU – as the Air Force preferred to call it) would be able to fly as an *independent* satellite. In August 1964, the Air Force awarded a contract to AiResearch in Los Angeles to incorporate the EVA Life Support System that NASA was developing for use as a chestpack into a backpack that had a system of thrusters. The development of the AMU was managed by the Air Propulsion Laboratory at Wright–Patterson Air Force Base in Ohio, and fabricated by the Chance Vought Company (later Ling Tempco Vought) in Dallas, Texas. At $12 million, this was the single most expensive Air Force experiment in the Gemini Program. When tests revealed that the 700 °C efflux from the hydrogen peroxide thrusters damaged the mylar insulation of the G4C suit, the exposed sections were reinforced with eleven layers of aluminised polyamide film as a thermal shield, and the legs were further protected by a woven stainless-steel cloth. The Air Force wanted the AMU to be tested on two missions. Gemini 8 was the first that could have accommodated it, but NASA gave priority to its own backpack-augmented HHMU and so Gemini 9 was given the AMU test. Gene Cernan and Buzz Aldrin trained to use it. The Air Force asked for Cernan to fly freely, but NASA insisted that he be tethered in case of an AMU problem. Given the manner in which crews rotated from backup to prime, the second test would be on Gemini 12, the final mission of the Program. If all went well on the first test, Aldrin might be authorised to fly without a tether, as was to be done on the Air Force missions.[2]

The Gemini 9 patch looked forward to a spacewalk alongside a docked Agena.

Atlas failure

At 10:12 Cape time on Tuesday, 17 May 1966, an Atlas left Pad 14 with GATV-5004. Ten seconds before the Atlas was to shed its two side-mounted engines, one of these engines gimballed hard over and forced the vehicle into a nose dive that dumped the Agena into the Atlantic. Guenter Wendt's Pad 19 crew retrieved a frustrated Stafford and Cernan from their spacecraft. The delivery of the next GATV was two months off. To reassign it to Gemini 9 would not only delay the Program, the fact that only six modified Agenas had been ordered meant that a later mission would be denied a target. In fact, as there was a risk of the final few missions being deleted if they could not be flown before the 'shakedown cruise' of the Apollo mothership, which the optimists hoped would be in late 1966, the Program could not accommodate another lengthy hiatus. However, NASA had an 'ace' up its sleeve, in the form of the Augmented Target Docking Adapter (ATDA) that had been fabricated in the aftermath of the loss of GATV-5002 for precisely such a contingency. Although the ATDA did not have the Agena's propulsion system, and therefore would be unable to manoeuvre following its release by the Atlas, the fact that the vehicle, at 2,400 pounds, was only one-third of the mass of the GATV configuration of the Agena meant that an Atlas would be able to insert it directly into the high circular orbit required to serve as a target in a Gemini rendezvous. While the ATDA would enable the docking to proceed, docked manoeuvres would not be possible, but this did not matter because no such manoeuvres were planned for Gemini 9.

In addition to cobbling together the ATDA, Charles Mathews had placed an Atlas on a 14-day requisition notice (in fact, this was the Atlas that was bought to launch

[2] The Air Force's plans for 'Blue Gemini' missions came to nothing, and its Manned Orbiting Laboratory was cancelled in 1969.

GATV-5001, and then placed in storage when it was decided not to launch that vehicle). In invoking this requisition on 18 May, he ordered that the Atlas be ready for launch by the end of the month. The only obstacle was the requirement to determine the reason for the recent Atlas failure. Fortunately, an analysis of the telemetry identified a short circuit in the autopilot, which was straightforward to rectify. On the same day, Mathews ordered that if the ATDA could not be placed into the required orbit, Gemini 9 would be inserted into an orbit that would enable it to rendezvous with the Agena left by Gemini 8. Although *that* vehicle's radar transponder was inert because the battery had long since expired, it would be able to serve as a passive target for the optical rendezvous tests. However, this plan was foiled when it was realised that the storage orbit was too high for the Gemini spacecraft to reach without employing an Agena's engine. On 31 May, therefore, Robert Seamans ordered that if the ATDA was lost Gemini 9 would be launched on time to enable Cernan to test the Air Force's AMU because the Program could not accommodate any further delay.

The stubby ATDA on its Atlas launch vehicle is prepared on Pad 14.

Frustration

The Atlas lifted off at 10:00 Cape time on Tuesday, 1 June and, to the delight of Stafford and Cernan, who were in their spacecraft on Pad 19, placed the ATDA into a perfect circular orbit at 161 nautical miles. The telemetry relayed to Houston by the Bermuda tracking station perplexed Jim Saultz at the Agena console, however. "Flight, I think we've got some problem with the ATDA," he called. "We're using attitude control fuel like crazy." The vehicle was evidently in trouble. "Also, I didn't see telemetry indications of the nose shroud separation." In fact, there were contradictory signs. The command to jettison the aerodynamic shroud that had protected the TDA during the ascent had been issued, but the signal to confirm that it had released had not been received.

As Saultz pondered whether the shroud had separated, the monitoring station on Canary Island picked up the ATDA's telemetry. "Flight," the CapCom called, "we're really hosing out the fuel." In fact, one of the two 'rings' was already exhausted. "I recommend that we secure the attitude control jets before we lose it all."

"Go ahead, shut it off," Gene Kranz concurred. Then he called Saultz on the loop, "Jim, keep me advised of any further developments. I'm going to follow the Titan from now on." Seamans's edict was clear – to launch Gemini 9 regardless. If Stafford and Cernan were able to rendezvous with the ATDA, they would be able to report on its condition.

The Titan countdown was halted at T–3 minutes to enable Carnarvon's tracking of the ATDA to be integrated into the computation of the best time to launch, and then, 98 minutes after the Atlas lifted off, the clock was resumed. At T–100 seconds, the Pad 19 blockhouse sent a final azimuthal update to the launch vehicle to tell the guidance system of its second stage how to steer into the target's orbital plane, but this signal was not received. While the controllers tried to identify the fault, the launch window expired. Each second after the best time to launch would make the Titan's task more difficult, and a 40-second delay would open the angle beyond that which the spacecraft itself would be able to correct, and once this time had elapsed the count was scrubbed. The problem was traced to a transmitter on the ground, which was soon repaired, but by that time Stafford and Cernan had once again descended the elevator for a two-day recycle. Meanwhile, the technicians who had prepared the ATDA for launch had reached the conclusion that the safing pin on the restraining strap around the top of the shroud had not been removed, with the result that even though the jettison sequence had initiated, the strap had failed to separate and had fouled the partially opened shroud. Commands were sent to extend and retract the collar of the TDA in an effort to shake off the shroud, but to no avail.

As Tom Stafford and Gene Cernan arrived in the White Room they found their spacecraft adorned with messages.

Finally off

On Friday, 3 June, after yet another steak and eggs breakfast, Stafford and Cernan made their way to the suit-up van on Pad 16. Because Stafford had had so many scrubbed launches, he had gained the moniker 'The Elevator Astronaut', and Cernan had come to the conclusion that his Command Pilot was jinxed. Their backups had strung a banner between the open spacecraft hatches which read, 'We were kidding before, but not anymore' and 'Get yourself into space, or we'll take your place'. Not to be outdone, Pad Leader Guenter Wendt presented Stafford with a 3-

foot-long match with which to sneak out and ignite the Titan himself in the event that the blockhouse called another scrub.

"I think we're going to make it, Gene!" Stafford observed to Cernan as the launch vehicle transferred to internal power.

As previously, the final navigational update failed but the blockhouse ignored the fault and pushed on. If it was launched on time, the spacecraft would be able to make up for any out-of-plane error inherited from the Titan.

"Okay, TP," Cernan prompted, eager to start the chase, "let's get him."

"Ignition!" Aldrin called at 08:39 Cape time. After preparing the spacecraft, he had gone to the blockhouse to serve as the 'Stoney' CapCom.

"Lift-off!" Neil Armstrong called. As commander of the previous mission, he was serving as CapCom in Houston for his successor.

"We're on our way!" Stafford reported delightedly.

"Beautiful, Tom!" Cernan exclaimed. "We're going."

The first stage's performance was nominal, but as they headed out over the Atlantic the early morning sunlight in the windows was so bright that they could not read the instruments. "We have the Sun in our eyes" Stafford warned Armstrong. "Keep us closely advised, Neil." Despite the building g-force, he raised his arm in an effort to shield his eyes.

In separating from the spent stage, Stafford made the IVAR burn to correct the launch vehicle's in-plane velocity errors.

"It was quite a ride, wasn't it?" Stafford, riding his second Titan, prompted Cernan, who was doing so for the first time.

Gemini 9 lifts off.

"Pretty fantastic!" Cernan agreed.

The Eastern Test Range tracked the spacecraft after the IVAR burn and Houston made a preliminary calculation of the phase adjustment to be made at apogee, a minute before making contact with Carnarvon on the western coast of Australia, and this data was read up by the Canary CapCom. As Gemini 9 passed over Africa, Armstrong called via Kano in Nigeria and revised this prediction based on Canary's data, so as they set out over the Indian Ocean they were in good shape.

"Hello Carnarvon," Stafford called.

"Go ahead," acknowledged Bill Garvin.

"We completed our burn right on time."

The phasing burn had raised the perigee from 87 to 125 nautical miles. Gemini 9 was now 425 nautical miles behind the ATDA.

"You look real good," reported Gary Scott when they flew within range of Hawaii several minutes later. "We've turned the L-Band of the ATDA 'On', and it looks good."

"Real fine," Stafford acknowledged.

"The manoeuvre plan that we're working on, Tom," Armstrong said, when the spacecraft flew into range of the relay site at Point Arguello in California, "has got a delta-H of 12 miles and TPI is 4 minutes before sunset." This slow approach would enable them to spot their target just before sunset and then maintain its acquisition light fixed against the stars during the terminal phase.

"Sounds real good," Stafford agreed. Of course, these times would vary as they made the two manoeuvres to establish the terminal phase, but this initial prediction was encouraging.

"There'll be no stars for either burn," Armstrong warned, referring to the two forthcoming set-up manoeuvres.

"Say again?"

"One burn is in daylight," Armstrong explained, "and the other is pointing down." The Flight Dynamics Officer had decided to correct the out-of-plane error in two stages, with each set-up burn incorporating a yaw component. The first burn would establish a 'node' at which to make the second burn where, in addition to other refinements, Gemini 9 would slide into the orbital plane of its target. But as Armstrong had noted, there would be no stars to provide a visual reference because one burn would be in daylight and the spacecraft's nose would be aimed downward for the other. After giving the data for these burns, Armstrong said, "You're on your own now."

"We're all set to go, Neil," Stafford assured.

Leaving the Eastern Test Range behind, Gemini 9 headed out across the Atlantic for the second time. In addition to refining the apogee precisely 12 nautical miles below the ATDA's orbit, the 'corrective combination' burn made ten minutes later served as the first part of the out-of-plane correction. On establishing contact with Houston via Tananarive on the island of Madagascar they reported their success, then prepared for the co-elliptic burn at apogee that would raise the perigee to circularise their orbit.

"You're about 150 miles. Is the radar light blinking yet?" Armstrong asked when he next established contact. As the radar picked up stray reflections prior to being able to lock onto the ATDA's transponder, its signal-acquisition light would flicker. The radar was designed to lock-on at a range of 200 nautical miles, and the fact that it was still not getting a return from the transponder was ominous.

"Not yet, but we'll give you a call when it does," Stafford promised. One minute later, he was back with, "There appears to be radar lock-on." The fact that it was intermittent implied that the ATDA's shroud had not been fully released.

As they closed within 130 miles, the light flickered and then went dark. "We've lost radar lock completely." Stafford reported. The light flicked for a moment, and then went off again.

"We're coming up for the burn," Cernan reminded Stafford. Trouble-shooting their radar would have to wait until they were safely in the catch-up orbit.

"Gemini 9, Carnarvon CapCom, standing by," Garvin called as the spacecraft rose above his horizon. He did not wish to distract the astronauts as they double-checked that they were set for the burn.

"We should be burning shortly," Stafford advised.

"Mark! Burn!" Cernan called.

"Flight," Garvin called over the network having monitored the telemetry on both vehicles, "the radar is still in-and-out."

"Can you give me any range or range-rate readings," asked Kranz.

"She's not locked on," Garvin pointed out.

"The burn's complete," Cernan reported. They were now in a co-elliptic orbit 12 nautical miles below that of their target, which was now just over 100 nautical miles ahead.

"We're not having too much luck with this radar, Flight," Garvin continued.

"Okay," Kranz acceded. If the radar's lock-on was intermittent, there was nothing anyone could do about it.

On contacting Hawaii, Stafford reported that they now had a solid radar lock but a serious problem had developed. "We've had a computer malfunction and we want to 'clue you in' on it."

"Go ahead," said CapCom Gary Scott.

"After that last burn, every time we switched from 'Rendezvous' back to 'Catch Up', the 'Comp Light' would come on without starting the computation and the IVIs would display." The switch settings were computer modes. The light came on when the computer was making a calculation. The velocity components for a burn were shown by the Incremental Velocity Indicators. Stafford gave a detailed description of the trouble-shooting process they had used to diagnose the fault. As they pursued their co-elliptic orbit slowly closing on the ATDA, the rendezvous procedure was to switch back and forth between modes. Every 100 seconds, the computer was to sample the angle of elevation and range and range-rate data from the radar until it had three data points, at which time it was to calculate an initial 'solution' for the TPI burn, which it would update as additional data points came in. However, it seemed as if every time that they switched modes the computer *restarted* the sequence, just as if they had hit 'Start Comp'.

"Houston says they will talk to you about the problem over the States," Scott replied. In trying to make a fast rendezvous, with time at a premium, the last thing that they needed was a computer failure. "The ATDA has been configured for rendezvous and all its lights are on."

"Real fine," Stafford replied.

As Gemini 9 slowly caught up, the elevation of the ATDA progressively increased. At a range of 75 nautical miles, it was 8 degrees above the local horizon and Cernan began to take sightings. As they started across the United States, Armstrong gave Houston's verdict on the computer. "We're going to work out the best time for you to cycle to 'Rendezvous', so that you get a correct solution at the proper time." That is, if the computer was able to make only one attempt at the solution, they would start it so that it would deliver its result just before it was required. The pitfall, of course, was that a single bad data point would ruin the prediction for the TPI manoeuvre, and with it the rendezvous.

"We've already thought of that," Stafford noted.

"I'm sure you did!" Armstrong agreed.

"Our analysis," Stafford said, "is that something *is* initiating the Start Comp cycle."

"You've got less than a mile of relative ellipticity," Armstrong advised, referring to the difference in the shapes of the two *nearly* circular orbits, "so it's going to be reasonably easy to predict the proper TPI time." In other words, they ought not to be overly concerned about the computer fault.

"We've completed the platform alignment," Stafford reported, "and are pitching up to the target." From now on, Stafford would keep his spacecraft's nose facing the target, and Cernan would plot the increasing pitch angle and the radar data on his chart.

"Our estimate for the time to switch to 'Rendezvous' is at 15.3 degrees," said Armstrong a few minutes later when he gave them Houston's advice for how to deal with the computer. The estimate for the time of the TPI burn was five minutes before sunset, which was perfect.

"That's pretty close to what we've come up with," Stafford replied. If they were lucky, the computer would run through the 10-minute procedure for the 'closed loop' solution. Two minutes later, he was able to announce, "I've got a real faint light out there by reticle, but it's still too early to tell."

"Good show!" congratulated Armstrong. At this point, he was joined at the CapCom's console by Slayton and Lovell and Aldrin, the backup crew.

A moment later, Stafford's uncertainty evaporated. "I've got it! I've got him in reflected sunlight, about a sixth magnitude star." With that, Gemini 9 headed out across the Atlantic to start its third revolution.

The computer finally delivered its result. The forward component agreed with Houston's recommendation, but the other components did not.

"You want another point?" Stafford asked Cernan.

Cernan waited and integrated another point, but the solution did not change significantly. With the angle already up to 25 degrees there was no more time, so Stafford decided to adopt Houston's solution.

"Exactly on time!" Cernan observed, when Stafford fired the thrusters. "Hey! I see it," he called a minute later.

"Oh, yeah, it's out there, man," Stafford agreed. "It's really out there."

"I hope that's not the shroud," Cernan said, wondering whether the dot of light was so bright because it was sunlight reflecting off the white shroud.

"We're getting dark," Stafford observed as they approached the terminator. "It's going to disappear shortly. It's getting dim – very dim." The target flew into the Earth's shadow and disappeared. "And I have the flashing light!" he said once his eyes adapted to the darkness. This was good news, because the acquisition lights were on the part of the vehicle which had been covered by the shroud, which suggested, despite the intermittent radar lock-on, that the shroud had indeed separated.

"You have?"

"I've got the acquisition lights," Stafford announced on the radio, but they were between stations and there was no response.

"I've got 'em too," Cernan confirmed a minute later after spotting the lights for himself.

"We're in business," Stafford said in satisfaction. But seconds later "I've lost them!" The lights had disappeared. "There they are again!" he exclaimed when the lights reappeared half a minute later. So, not only was the shroud still partially attached, their view of the acquisition lights was being periodically blocked because the vehicle was unstable.

There are no stars out there," Stafford complained, some 20 minutes into the transfer and with only 5 nautical miles to go. He was to maintain the target fixed against the stars, but he couldn't because moonlight was interfering with his dark adaptation. If they had been able to launch on 1 June as planned, then the Moon would not have been in his field of view at this juncture. "I'm tracking him on radar." The radar needles were not as accurate as the reticle, but he had no other option.

"It looks fairly good, Tom," Cernan opined as he plotted their approach on his chart. "As a matter of fact, it's looking *real* good!"

"We've got to get in there *somehow*, Gene."

Ironically, as the pitch angle approached 80 degrees, they finally got a clear view of their quarry by the moonlight that it reflected.

When Garvin in Carnarvon announced that he was standing by, Stafford said that they were pitched up at 88 degrees and had just over 3 nautical miles to go. "We're coming up the pipe!"

"Good show!"

"It looks like we're going to be braking directly into the Moon, Gene," Stafford noted. "It should be an interesting problem."

"Too fast," Cernan warned, as they closed within 2 nautical miles at 37 feet per second. He was monitoring their rate of approach. "Way too fast, Tom." But then he realised that he had made a mistake. "No, that's about right. I'm sorry. That's exactly right!"

By now they could see the ATDA's acquisition lights again and were sufficiently close to resolve the angular separation between them. "It's tumbling," Stafford concluded. Even if the shroud had separated, they would not be able to dock with the vehicle unless it could stabilise itself.

"110 degrees, 1 mile, 23 feet per second," Cernan reported just before they flew below Carnarvon's horizon.

"Should we brake it down some?" Stafford asked.

"Yeah," Cernan agreed.

Stafford could now see *one* star in his reticle, so he endeavoured to hold the ATDA fixed with respect to this as they passed through 2,000 feet at 20 feet per second. At 1,500 feet, he slowed to 7 feet per second, and maintained this rate. Contrary to expectation, the vehicle was fairly stable – it was rolling, but not tumbling. By the time they closed within 600 feet, Gemini 9 was pitched up at 130 degrees – which

To convey an impression of how the aerodynamic shroud was fouled, Tom Stafford likened the vehicle to "an angry alligator".

would be its maximum angle – and Stafford and Cernan completed the rendezvous flying almost heads down.

"Look at that moose!" Stafford exclaimed as the vehicle's configuration became apparent in the moonlight.

"I wish we had daylight, to see this thing better," Cernan complained.

"I've got to stop," Stafford decided, at 360 feet. As he slowed, he continued to inspect the target, endeavouring to determine its condition. "The shroud is *half open* on that thing!"

"It sure is," Cernan agreed.

"The back pin didn't fire?" Stafford speculated.

"It's like a clam shell, Tom."

Passing through 240 feet, they inspected the target in the beam of their docking light.

"Look at that thing!" Cernan exclaimed upon realising how precariously the halves of the shroud were clinging to the collar of the TDA. "You could almost knock it off! I feel certain you could. The band didn't break, I think. What a mess!"

Fly-around inspection

In the expectation that Gemini 9 would be station-keeping with the 12-foot-long ATDA by the time Hawaii acquired them, Kranz asked Gary Scott to ask Stafford how badly it was tumbling and, assuming that the shroud had released, whether he thought the docking would be feasible. But the news was not good. "We've got a weird looking machine here," Stafford reported.

"What's it look like?" Scott enquired.

"Both the clam shells on the nose cone are still on," Stafford began a commentary for the engineers. "The front-release has let go. The back explosive bolts – attached to the ATDA – have both fired and it appears that one of the bolts from the band has fired. What's keeping it together, is the 'quick disconnect' for the small electrical connector that fires the bolts on the band."

"Understand."

"The shroud's like an alligator jaw that's open about 25 to 30 degrees," Stafford mused, giving the journalists a vivid analogy, and the headline for their commentaries. "Both of the piston springs look like they're fully extended. Also, the back parts of the nose cone have separated from the ATDA – it looks like it's just held on by some inconceivable force." The spring-loaded pistons had attempted to force the halves of the shroud apart, but the band that was constraining them had prevented their base from clearing the outer rim of the conical docking collar. "But everything else looks good," he concluded wistfully.

"How bad is it tumbling?" Scott asked.

"It's 3 or 4 degrees per second." If they could detach the shroud and the ATDA stabilised itself, then they would be able to attempt the docking.

"Ask the crew if they think cycling the TDA from unrigidised to rigidised might shake off the shroud," Kranz prompted Hawaii.

Stafford was making an inspection fly-around of the ATDA. "The basic rate is in roll," he decided. "The body axis is nearly horizontal." The configuration of the shroud had evidently confused the Attitude Control System, and it had not been able to properly orient the vehicle.

"Do you think there's any possibility of breaking the cone – shaking it loose – by going through a few unrigidising-rigidising sequences?" Scott asked.

"Well I might give it a try," Stafford allowed.

At this point, Kranz seized the initiative. "Why don't you have them stand by there, and you try cycling the TDA cone to rigidised and then back to unrigidised, and have them watch it to see if it seems to be doing any good."

However, Stafford had now drawn up to within 3 feet of the vehicle to describe the fouled mechanism in more detail.

"We'd like to go through one rigidising and unrigidising sequence," Scott called, "if you'll get into position where you can watch it."

"Standby one, I don't want to be too close when that thing cuts loose," replied Stafford. He then resumed his inspection commentary, concluding: "It looks like someone didn't hook up the disconnect cable." The sequence was clear: "The disconnect didn't disconnect. That's the only thing that's holding the whole mess intact." His verdict: "It's basically free of the ATDA. It's just barely held on by those four pyro wires in there – those little wires on the strap." His recommendation: "We might put out our docking bar and go up and tap it." He wanted to extend a short bar built into the nose of his spacecraft and nudge this against the shroud to try to shove the base of one of its clam shell segments over the rim of the docking collar, in the hope that with one half free the entire thing would slip off.

"Standby," Scott warned.

"Tell them we're going to work on their description," Kranz told Scott, "and we'll cycle the TDA when we pick them up over the States." He was not enthusiastic about letting them nudge the shroud. He did not want to risk damaging the radar, the re-entry thrusters, and the parachute, all of which were within the spacecraft's narrow nose section.

In the final moments of the Hawaii pass, Scott relayed the Flight Director's instructions, and Stafford said he would withdraw to a safe distance and wait.

"We're standing by for an unrigidising sequence," Stafford said as they started across the US. "If we could get it right now, it would be great."

"We'll have to wait until you get in Canaveral's acquisition," Armstrong pointed out.

"Have you got a reading on my suggestion for extending the docking bar and giving it a tap?" Stafford asked. "The whole thing may fall apart."

After Stafford had described how the shroud had been fouled, an impromptu meeting in Mission Control discussed having Cernan swing open his hatch, stand on his seat, and try to release the shroud manually. Aldrin, Cernan's backup, was in favour. But with the vehicles that close there was a risk of a collision damaging the hatch. Charles Mathews, the Program Manager, suggested that Cernan might plug in his long umbilical and jump off towards the ATDA, grab hold of it, and then reach inside to pull a 'quick release' lanyard or, if necessary, use scissors to cut the metal strap that was holding the two sections of the clam shell against the docking collar. However, cutting the strap posed a serious risk. The casing had been split lengthwise by pyrotechnics, but the two springs – whose action was being inhibited by the strap – delivered a force of 300 pounds each, and when Cernan cut the strap they would eject the shroud segments, whose jagged edges might rip his suit. The debate was still in progress, and Armstrong did not know which way it was going, so he told Stafford that they would hold his "contingency plan 'B'" in reserve. However, he did know what the Agena engineers had concluded had gone awry during launch preparations. "We're pretty convinced – due to the telemetry signals from the cables – they didn't pull the disconnects to the wire bundles, they're still plugged in."

"The cables might still be there," Stafford admitted, "but once that strap goes, the whole thing should deploy."

"We're ready to send the commands now," Armstrong called as they approached Florida, referring to the plan to cycle the TDA.

"We're right behind it," Stafford advised. By positioning Gemini 9 behind the ATDA's engine, they would be clear of the shroud if it separated. "It's moving all around." The motion of the collar animated the alligator's jaws and pitched the vehicle, but the strap held the parts of the shroud in place. The attitude control thrusters fired as the vehicle attempted to recover from this disturbance. "It's not doing any good," Stafford concluded dejectedly. "You might as well save the fuel."

Second rendezvous

As Gemini 9 flew down the Eastern Test Range on its fourth revolution, Kranz ordered it to withdraw in order to proceed with the second rendezvous, while the engineers mulled over the state of the ATDA. So it was that barely 45 minutes after drawing up alongside its target, Gemini 9 prepared to depart. If the mission had unfolded as planned, this separation to start the 'equi-period orbit' rendezvous to determine whether the TPI burn could be calculated by optically tracking the target would have occurred at T+28 hours; Kranz had simply brought it forward a day.

The plan was to make a radial burn that would ease Gemini 9 into an elliptical orbit with the same orbital period as the target's circular orbit, but with its apogee, 90 degrees around from the separation point, some 2.5 miles above it, and its perigee, 180 degrees later, a similar distance below it. When Gemini 9 was above the ATDA, it would move more slowly and slip behind, and when it was lower it would catch up.

By the time Gemini 9 reached the opposite side of its orbit from the separation point, it would be trailing the target by 11 nautical miles. At perigee, Gemini 9 would be able to set up for TPI as if it had just made the co-elliptic burn – although because the altitude difference was only a few miles, the transfer would be shorter, subtending a central angle of 80 degrees instead of the canonical 130 degrees. Cernan was to employ a sextant to measure the elevation of the target above the local horizon, to enable the computer to calculate the time of the TPI burn for a rendezvous shortly before orbital sunset.

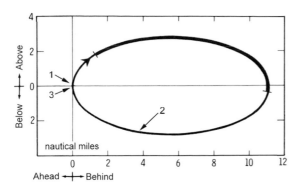

How Gemini 9 re-rendezvoused with the ATDA from an equi-period orbit, with the terminal phase performed in daylight.

1. Radial separation burn (the thick section of the loop indicates darkness)
2. TPI 1h 13m later for an 80-degree transfer
3. Interception after 1h 33m

As their prospects were strongly dependent on the 'purity' of the radial separation burn, Stafford aimed his spacecraft's nose directly at the Earth, and then took up station above the ATDA and nulled out his relative motion, finally declaring himself to be satisfied with barely a minute to spare.

"We're standing by for your manoeuvre," announced Keith Kundel, the CapCom aboard the *Rose Knot Victor*, which was stationed in the South Atlantic.

"Mark the burn!" Stafford announced. The forward-firing thrusters were used to impart a purely radial impulse.

Sunset was 8 minutes after separation, so while they still had sunlight Cernan lined up the sextant and tried to track the ATDA as it slipped away beneath them, and found that keeping track of the bright speck moving against the backdrop of clouds was extremely difficult. After sunset, however, he was able to follow its acquisition lights against the dark planet below.

"Our trajectory looks real nominal," Stafford reported to Houston through Tananarive.

"Very good," replied Armstrong.

To check on their progress Cernan took note of when the target passed through the local horizontal, which was within 20 seconds of the nominal time.

The rotation of the Earth meant that on this pass the spacecraft did not come within range of Carnarvon, but the *Coastal Sentry Quebec* picked them up as their track turned northeast across the Pacific.

"We're exactly nominal," Stafford reported.

Soon afterwards, they flew into daylight again, and into Hawaii's range. "What about the ATDA?" Stafford asked.

"We haven't got any more word on it," Gary Scott reported.

As they began to catch up with the ATDA, Cernan started to take sextant sightings, but 'airglow' made the horizon indistinct, complicating measuring the vehicle's angle of elevation. When the computer ran the calculation, it produced a surprising result. On checking, Cernan found that he had miskeyed the data. "I fouled up! Sorry, Tom. I'll get it. I'll get it."

In daylight the ATDA shone as brilliantly as Venus, so Stafford rolled inverted to put the Sun behind the spacecraft's nose in order to enable him to view the target without his reticle being flooded with light.

"We're approaching the terminal phase," Stafford confirmed, when Armstrong called one minute before the nominal time to TPI to say that he was standing by via the Corpus Christi station in Texas. When the onboard computer completed its calculation, they found that the time for the burn had just past, so Stafford made it straightaway.

"We're with you," Armstrong assured, when he heard that they had started the transfer a little late.

"Everything's looking good," Stafford reported optimistically. This time, since they could not use the radar, there was no need to maintain the spacecraft's nose facing the target, indeed it had to remain horizontal so that Cernan could measure the target's elevation by the sextant.

"I hope we're not overshooting it," Cernan said as they caught up with the target.

"What do you think of the sextant?" Stafford enquired.

"Well, it's alright," Cernan allowed, "but I've got such limited space to work with." In the post-flight debriefing he recalled the procedure: "It takes a minute or so to get that sextant set to take a sighting. To acquire the target, to pick up the horizon, to take the measurement, to take the sextant down off the window mount, to put it under the lamp, and to read its angle is a time-consuming operation." Furthermore, the small size of the window prevented the sextant measuring angles higher than about 70 degrees when it was on its mount. When he held it manually, the fact that he could see only a short arc of the horizon made it difficult to ensure that his measurements were perpendicular. He had proved the concept, but the device was too bulky for routine use in the cramped confines of the Gemini cabin. Perhaps in the much larger Apollo spacecraft it would be more practicable.

"I'm on him," Stafford confirmed, as they closed within a mile of the target vehicle at a speed of 19 feet per second.

"Are you accurate enough for line-of-sight?" Cernan asked.

Stafford reckoned that he could judge the range from the size of the target in his reticle. "It's really starting to grow."

"You're inside a half-mile," Cernan noted. "You'd better start braking." Stafford did so, but as they passed through 2,000 feet at 11 feet per second Cernan urged further braking. As they flew directly below the target he realised that his final handheld sextant sighting had been bad. "I really screwed up on that one, Tom. I'm sorry!"

"We had a good transfer, though," Stafford consoled. "I'd estimate that we're 1,000 feet." On passing within 600 feet, he slowed to 3 feet per second. As they pulled up alongside the ATDA for the second time, the *Rose Knot Victor* announced its presence.

"We're there!" Stafford reported. "We braked optically, and I'd estimate that we're about 200 feet out and closing in slowly."

"Roger," Kundel replied matter-of-factly. Then he delivered the bad news: "I have some information for you. The people at the Cape and Houston don't believe that we can get the shroud separated."

"Okay," Stafford replied, accepting that they had lost their docking. But docking was not new and, thinking positively, they had accomplished a demanding optical rendezvous. A few minutes later, they flew into the Earth's shadow.

In the post-flight debriefing, Cernan said: "I think that this rendezvous showed the way that the lighting conditions should be for a passive target on an optical rendezvous." This was good news for the Gemini 10 crew, who were planning to undertake such an approach to the Agena left by Gemini 8.

Off again!

Moving on, Keith Kundel relayed Houston's latest instructions: "They want you to use minimum OAMS during your station-keeping period. If you have 40 per cent, we'll give you the updated flight plan." As the PQI was 42 per cent, the third rendezvous was 'on'. Half an hour after drawing alongside the ATDA, they were off again.

This time the separation was to be a 6-second retrograde burn with the spacecraft SEF and using the forward-firing thrusters. "Remind him to do it when he's *behind* the ATDA," prompted Glynn Lunney, who had taken over as Flight Director. The manoeuvre would slow Gemini 9's motion in its orbit, and Houston wanted to ensure that there was no possibility of a collision. It could have been made using the aft-firing thrusters, but flying BEF would deny the crew a view of the other vehicle as they withdrew.

Tom Stafford inside.

On establishing contact with the *Coastal Sentry Quebec*, they reported that the burn had been made, and the PQI was 40 per cent.

"How's it going?" Gary Scott asked when they reached Hawaii.

"We're powering down," Stafford reported. After a bite to eat at the end of a long and eventful day, they were to start their first sleep period.

Half an hour later, Kundel sounded them out about the prospects for a rendezvous from above in daylight using optical tracking.

"We can rendezvous in just about any type of situation you give us!" Stafford assured. He now had more rendezvous experience than anyone else – having made one on Gemini 6, and two thus far on this mission.

Gene Cernan inside.

By the time of Armstrong's wake-up call eight hours later, Gemini 9 was 60 miles *ahead of* its target. Without much of a preamble, he gave them the data for the first two manoeuvres, the first of which was due in 30 minutes. A 2 foot per second burn was to refine the phasing for the 17 foot per second height adjustment 180 degrees later that would establish an apogee point 7 nautical miles above the ATDA's orbit. "Be advised," he added, "that they've got a shroud plan in work."

"Okay, fine!" Evidently, contrary to what had been said the previous day, the docking had not been written off after all.

As the ATDA's radar transponder had been activated, and Gemini 9's radar was tracking it, Armstrong advised them to orient the spacecraft BEF and use the forward-firing thrusters for the co-elliptic manoeuvre that was to raise the perigee and circularise their orbit 7 nautical miles above that of the ATDA, "so that y'all won't have to turn around and lose radar lock."

"Thank you," Stafford acknowledged.

The burn was completed without incident.

"How're things in Houston this morning?" Stafford asked casually as they started across the United States on the next pass.

"Oh," replied Armstrong, "we've been as busy as beavers down here."

"I can imagine!"

"You guys keep terrible hours," Cernan commiserated; it was not yet 06:00 in Houston.

"You do too," Armstrong observed.

After Gemini 8, Dave Scott had been assigned as Senior Pilot on Jim McDivitt's Apollo crew. When Gemini 9 launched, they were at the North American Aviation plant at Downey near Los Angeles. After the discussion in Mission Control concerning whether Cernan could manually release the ATDA's shroud, Slayton sent them over to the Douglas Aircraft factory at nearby Long Beach to inspect a shroud and report on the prospects for effecting a manual release. For fidelity, they tracked down a pair of surgical scissors similar to the ones carried on Gemini 9. "They've been climbing around a shroud out at Douglas for the last few hours," Armstrong reported, "and they advised that the outside isn't a problem, the inside may turn out to be a problem – there's quite a bit of sharp edges and things on the inside." Interrupting himself, he asked pointedly, "Did you happen to just punch 'Start Comp' or change mode or something in the computer?" Stafford said that he had been testing the computer. Armstrong said that it had shown up in the telemetry and he was just checking that they had not struck a switch by accident.

"It looks like our computer has cleared up completely," Stafford noted. The fault that had interfered with the first rendezvous was no longer present. This was good news. Even so, he was baffled. "I don't know what the glitch was."

Being in a lower orbit, the ATDA was catching up with Gemini 9. From the astronauts' perspective, of course, it was *they* that were approaching it. As the terminal phase was to be flown in daylight, the TPI burn would be some time near to sunrise. However, they could not see the ATDA's acquisition lights at all. Their first sighting of it was in reflected moonlight at a range of 20 nautical miles. Meanwhile, Cernan fed the radar data into the computer and ran the three-point computation.

"Here comes sunrise," Stafford noted. He was able to continue following the ATDA in reflected moonlight for a while against the dark side of the terminator but as it passed through the glowing horizon he lost it. "I'm flying on the radar needles now," he reported. Cernan did not like the computer's TPI solution, so he let it integrate a fourth point to get a better result, which Stafford decided to use rather than Houston's backup solution. "We're staying 'closed loop'," he announced.

"Understand," Armstrong acknowledged.

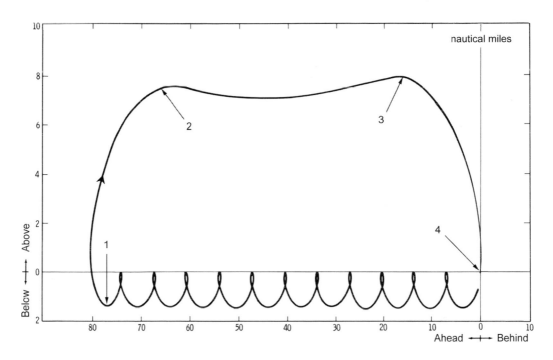

How Gemini 9 re-rendezvoused with the ATDA from above, tracking it optically against the Earth's surface.

1. Height adjust
2. Co-elliptic manoeuvre
3. TPI
4. TPF

As they approached the TPI point, Stafford maintained the nose of the spacecraft pitched *down* at the ATDA. During the early part of the transfer, they were over the North Atlantic. "We can't see him against the sunlit ocean," Stafford complained, "even though we're down to 12 miles range." Fortunately, the radar worked "beautifully" against the Earth background. "We're down to a little over 3 miles," he reported via Kano. "We're looking straight down at the Sahara – there's no visual contact." The transfer had been arranged specifically to enable them to view their target against the dunes and dark lava flows of the desert, the bleakness of which, it was believed, would be a fair approximation of the Moon's surface. They had it on radar, of course, but that was not the point – an Apollo mothership rescuing a lunar module stranded in low orbit would *not* have a radar.

"Your fuel cut-off is 5 per cent," Armstrong warned.

Propellant was not an immediate concern to the crew, however, because they had plenty and the radar indicated that the rendezvous was progressing nicely. If only they could *see* the target! Stafford knew where it *ought* to be, because his reticle was boresighted with the radar. "I finally have a little spot down there!" he reported at just less than 3 nautical miles. It was difficult to hold the target, however, because the albedo of the terrain kept changing and the drift-rate across the field of view of his reticle was striking. As they set out across the Indian Ocean, he was readily able to track it as a white dot against the blue water, but lost it while over cloud formations. He passed through 1 nautical mile at 13 feet per second, which he felt was "a comfortable rate", then started braking. However, as the ATDA loomed in his reticle he decided that he was coming in too fast and in over-braking attained a *negative* range-rate, which he then had to correct. Nevertheless, when Carnarvon picked them up, he was able to report that they had completed the rendezvous, and were again station-keeping. The PQI was 18 per cent. Garvin told them to remain with the ATDA, because when they reached the States Houston would brief them on their next task, so they had their breakfast as they flew across the Pacific.

"We're pretty bushed," Stafford pointed out via Canton, turning to the prospect of the scheduled EVA. "We've been busier than one-handed paper-hangers up here. Geno and I talked it over, and we think we should knock it off for a while. Perhaps we should wait until tomorrow morning."

"Roger, Tom," Armstrong acknowledged just before the spacecraft flew out of range. "I followed that." He came back when they reached Guaymas. "It's the ground recommendation that we postpone the EVA activity until the third day."

"We agree very heartily with that recommendation," Stafford confirmed.

Houston was not yet prepared to abandon the docking target, and while it was over the United States they tried to shake the shroud off by cycling the docking cone while the thrusters pitched and rolled the vehicle. Stafford withdrew 100 feet to observe. Although dramatic to observe, the effort was to no avail.

"Were you able to find out, Neil," Cernan enquired afterwards, "whether there's anything we could do during EVA to the ATDA that might afford us a docking?"

"We looked at it rather extensively," Armstrong replied. "We had some possible actions. However, we didn't have a very high confidence level in those actions." This was something of an understatement. In truth, the procedures were rather vague and Cernan's actions would have depended what he found when he examined the shroud, and the advice on offer was really only a list of objects that he was to avoid touching.

If the ATDA's gyrations had shaken off the shroud, Gemini 9 would have moved in and docked, but the Flight Director decided to cancel this objective. The mission was barely 23 hours old but it had accomplished a triple rendezvous, which was a great achievement by any measure. Although the from-above-in-daylight rendezvous had been successful, this had been because Stafford had used the radar to boresight his reticle. "If it had not been for the radar," he said afterwards, he "would have blown that rendezvous". The fact that they had spotted the target visually only in the final few miles implied that the likelihood of successfully using this method was very low, which was not good news for the Apollo planners.

Stafford and Cernan were told to abandon the ATDA, make a separation burn, and leave their spacecraft in free drift to save propellant, and devote the remainder of the day to their assigned in-cabin experiments.

Robert Gilruth, George Low and Chris Kraft on completion of Gemini 9's third rendezvous.

Cernan's ordeal

On being awakened 45 hours into the flight, Stafford and Cernan had their breakfast and, about an hour later, began to prepare for Cernan's EVA. Three hours had been allotted. The checklist with hundreds of items occupied eleven pages of the flight plan. Houston urged him to 'tank up' on water, so as not to risk becoming dehydrated during the excursion, which was scheduled to last 2 hours and 40 minutes. As they had no requirement to communicate with the ground during this preparatory phase, the WWTN sites prefaced their occasional updates with an instruction not to acknowledge. In one of his few reports, Stafford told Canary Island that they had unpacked the 25-foot umbilical. "We've got the big snake out of the black box." As soon as the strap had been released, the coil had expanded to fill the cabin. After strapping on the EVA Life Support System (ELSS) chestpack, Cernan plugged in the umbilical to make a suit integrity test. They were right on the timeline.

But on making contact with Tananarive, Stafford announced, "I'm having a little trouble." A thruster was creating a 25 degree per second roll. "Yaw-right and roll-left seems to be our main problem." This posed a dilemma, because if it happened while Cernan was outside, the rolling spacecraft might 'wind up' the umbilical, incapacitating him. After a stuck thruster had jeopardised Gemini 8, the OAMS system had been modified to prevent faulty thrusters from 'failing on', so the failure was puzzling. As they flew over the Indian Ocean, Stafford further investigated. On contacting Carnarvon, he said that although he now had the spacecraft under control, he could not explain why the fault seemed to be dependent on the *mode* of the flight control system. At this point, Gerry Griffin at the Guidance, Navigation and Control console in Houston asked Armstrong to prompt Stafford to check the circuit breaker for the horizon scanner which, as Griffin had suspected, had popped out; re-engaging the breaker eliminated the problem. "Good show," Stafford praised. "We'd thought of about everything else but the breaker!" It had been disturbed as they unpacked apparatus for the EVA. "Sorry about that," he apologised.

"There'll be days like that," Armstrong consoled.

Cernan was to open the hatch at sunrise on the 31st revolution in order to have an entire daylight pass for the initial part of his program. In the first five minutes, while 'standing' on his seat with Stafford holding an ankle strap to stabilise him, he was to eject the trash, which included the packaging for the EVA apparatus, retrieve a micrometeoroid collector from the adapter immediately behind the hatch, which he would pass to Stafford for storage,[3] deploy a pop-out handrail running back along the 'top' of the adapter, affix a 16-mm movie camera to a mount on the adapter, facing forward to document some of his later activities, and then make his way to the nose of the spacecraft to affix a rear-view mirror to the docking bar to enable Stafford to monitor his progress when he ventured to the rear of the spacecraft.

In the second phase of the program, lasting 30 minutes, Cernan was to undertake mobility and stability trials in front of the camera, to evaluate his ability move purposefully by tugging and swinging his umbilical and by using velcro pads mounted on the surface of the spacecraft and on his gloves. Back at the hatch, he was to dismount the camera and pass it to Stafford to have the magazine replaced, then remount it. The feature of the EVA, the AMU trial, would start towards sunset. After swinging his hatch almost shut to protect its rubber seal from the chill as the spacecraft passed through the Earth's shadow, he was to make his way along a handrail to the rear of the adapter, where the AMU was stored surrounded by handrails and a footrest incorporating a pair of metal stirrups. Two lamps had been installed to enable him to see what he was doing in the darkness. Having stabilised himself with the toes of his boots in the stirrups facing the AMU, he was to check valves and confirm the pressure of the AMU's hydrogen peroxide propellant, and then rotate down its two control arms. If he was satisfied, he was to extract his feet from the stirrups, turn around and reinsert his toes into the stirrups and then settle himself between the arms of the backpack and strap it on. Because the field of view of his helmet was limited, he would have to don the backpack 'by feel'. After plugging the AMU's oxygen supply into his ELSS, he was to disconnect the umbilical. The next task was to verify the warning lights mounted on the top of the ELSS. There were six lights to alert him if the pressure in the hydrogen peroxide tank dropped; if the tank was running dry; if the pressure in the oxygen tank dropped; if there was a drop in pressure of his suit; if the power supply to the thrusters failed; and if there was a problem with the emergency oxygen supply in the chestpack. Each condition was

[3] This experiment had been opened and closed by remote control. In fact, postponing the EVA doubled the time available to the experiment, which was a bonus for the experimenter.

accompanied by an audio tone, which Stafford would also be able to hear. As a safety measure, Cernan was to connect himself to a 125-foot nylon tether that was attached to the rear of the adapter. At sunrise, if Cernan was satisfied, Stafford was to throw a switch to blow the explosive bolt that would release the AMU from its mount.

Once free, Cernan was to undertake a series of tests to verify that the AMU's jets, each of which delivered 2.3 pounds of thrust, produced 'pure' motions, and that its 'attitude hold' function worked. Although three-axis stabilised, it had jets only for forward–aft, up–down, and roll, pitch and yaw rotations; lateral motions had to be 'composed' by rotating, moving, halting and then rotating back into the initial orientation. For the trial, Cernan was to move 80 feet beyond the spacecraft's nose, and then, in view of the camera, trace the shape of a '4' by moving 80 feet off to one side, then diagonally up across the nose until he reached the end of the tether. With this demonstration over, he was to fly directly to the adapter, reconnect to the umbilical, disconnect the tether, doff the AMU and push it clear of the spacecraft. After making his way back along the handrail to the hatch, he was to swing it wide and resume his 'stand up' position in order to photograph the horizon 'airglow' phenomena that complicated the use of a sextant. Finally, he was to retrieve the movie camera and close the hatch just after sunset.

Given that Ed White's excursion had lasted only 20 minutes or so, and had not required performing any real work, Cernan's assignment was very ambitious – far more so than that assigned to Gemini 8. In fact, if they had not been obliged to discard the ATDA, Cernan was to have retrieved a micrometeoroid package similar to the one that Dave Scott was to have retrieved from his Agena. Nevertheless, Cernan had trained hard, rehearsing each task in brief periods of weightlessness on a KC-135 flying a ballistic arc and was confident that the tasks were feasible.

"We're going for a walk," Cernan delightedly announced when they flew within range of Canton Island. He twisted the large handle above his head to unlock the hatch, then swung it wide, stood on his seat and ejected the trash. They were still in darkness but sunrise was imminent. An orbital dawn was spectacular when viewed through the tiny cabin window, but the view out of Cernan's faceplate was stupendous. The Sun was already up when Ed White had opened his hatch, so Cernan was the first spacewalker to witness the full splendour of an orbital sunrise. "Boy, it sure is beautiful out here, Tom. If I had my Hasselblad, I'd take a picture." But he had things to do. Turning aft, he retrieved the micrometeoroid experiment and passed it to Stafford, verbally coordinating their actions to ensure that at no time was the package free and able to float away. His first task completed, Cernan reached back to release the pop-out handrail recessed into the surface of the adapter, but it was so far back that as he reached for it his feet lifted off the seat and he began to float out of the hatch. Stafford, in the process of stowing the package, grabbed one of his feet and pulled it back down onto the seat. It took longer than expected to deploy the handrail. When

Gemini 6 had rendezvoused with Gemini 7, each crew had reported long 'tapes' trailing from the rear of the others' spacecraft. "I don't see anything waving off the adapter, from here," Cernan reported. At least he would not have to worry about becoming entangled. Stafford passed him the Maurer movie camera. "It's a strange world out here, you know," Cernan mused as he struggled to install the camera on the bracket just behind his hatch. "I had it in, but it wouldn't snap down."

"Try it again."

"Pull me down."

A still from the movie of Gene Cernan wrestling with his umbilical during his spacewalk.

With improved leverage, Cernan tried again. "That was a hard fit, but it's in." Although these tasks had been straightforward in training, in space he had found them awkward. If he had leverage and one hand free to work, then it was easy, but if the task required the use of two hands his unstabilised body floated out of position when he tried to apply a force to a piece of apparatus. The preliminaries had consumed more time than scheduled, so he took "a breather" for several minutes. As they approached the US mainland, he could see all the way from San Francisco, south into Mexico. It was "like sitting on God's front porch", he would write later.

After his siesta, Cernan pulled out sufficient umbilical to enable him to reach the nose, to affix the mirror to the docking bar. Whenever he moved, the umbilical thrashed. "Boy, that snake's really running around out here!" With the mirror installed, he pushed himself off the spacecraft to start the umbilical-dynamics experiment, the objective of which was to find out whether it was practicable to control his mobility and stability by pulling on and twisting the umbilical. To his surprise, the slightest change in posture twisted the umbilical, which reacted and perturbed him in return. It did not become evident until later, but as he performed these gyrations, he tore open a seam in the back of his suit, with the result that his lower back got a severe sunburn through the parted layers of thermal insulation.

Stafford managed to snap a few pictures before Cernan – out of control – drifted beyond the field of view of his window. As Cernan tugged on the umbilical, the vehicle pitched down 40 degrees, yawed 150 degrees, and rolled inverted, but Stafford was reluctant to use the thrusters for fear that the hot efflux would contaminate and possibly damage Cernan's suit. "Let me know if you get near those thrusters," he urged.

"I don't know where I'm going, right now, so hold on a few seconds."

"Let me know when you get on the adapter."

"The snake's all over me!"

"What?"

"I'm going to try to get out of here by playing out the umbilical." On stretching it to its limit, Cernan rebounded and the umbilical curled up as it 'remembered' its stowed form, and each time it twisted it forced him in the opposite direction. On colliding with the spacecraft, he snapped off one of its antennas. He realised that to attempt to control his movement using the umbilical was impractical because it had a life of its own – it was "like wrestling with an octopus", he later wrote. "I can't get over to where I want to go," he complained. "I'm near the aft thrusters." Stafford hastily switched them off. As he passed low across the adapter, Cernan managed to grab the handrail, and was finally able to anchor himself. "I'm going to start using one of the velcro pads," he reported. He had one pad on each glove and there were others at a test site on the adapter. "It came right off in my hand!" He tried again on another pad. "Okay, I'm stuck on this thing with a hand pad right now, but it won't stay." He was weightless but he still had mass, and his inertia all too readily tore him free. "The velcro isn't strong enough," he decided. As he clawed his way along the length of the umbilical, he made a discovery. "The only control that I have is the umbilical – and the shorter it is, the better the control I have." However, if he did not take care to pull through his centre of mass, the induced rotations sent him tumbling. Some of his gyrations had been documented, so the EVA planners in Houston would be able to study his performance in detail. "I'm coming back to the hatch," he announced. "What time is it?"

"About 24 minutes after sunrise." They were passing over the southeast United States, pretty much on schedule. "Ease around here and take a rest." Then Stafford switched from the intercom to the air-to-ground channel. "Houston, Gemini 9."

"We're still with you, Tom," Armstrong replied immediately. In order not to disturb the crew, he had remained silent all the way across the continent.

"Gene's done the umbilical evaluation, and it looks pretty rubbery. I've got him behind the cockpit now, and can see him real well in the docking bar mirror."

After a few minutes rest, Cernan retrieved the Maurer so that Stafford could replace the film magazine, then he reinstalled it and, as they flew beyond the Eastern Test Range, swung the hatch down until it was ajar only a couple of inches. As he moved hand over hand along the handrail, he pinched his umbilical into small rings along the way to hold it tight against the vehicle after he had disconnected from it. Stafford threw switches to extend the hand and foot restraints in the rear of the adapter and jettison the AMU's thermal cover. On reaching the lip of the adapter, Cernan got a nasty surprise – the ring that had been connected to the booster was a sawtooth of very sharp-looking metal. He gingerly slipped over the rim into the cavity at the rear. Although this excursion had been straightforward in training, by the time Cernan reached the stowed AMU his heart was racing at 155 beats per minute and he was sweating heavily, so Stafford switched the environmental system to its 'High' flow rate to feed oxygen to cool him down. The schedule required Cernan to achieve this point by sunset over South Africa because the only illumination available while in the Earth's shadow would be the lights in the rear of the adapter, and then be ready to start the AMU test at sunrise. Although one of the lamps had failed, he slid his toes into the stirrups and pushed on. As an independent spacecraft with its own manoeuvring, life support and communications systems, the AMU was a complex apparatus, and he had to work through a 35-item checklist. The procedure was complicated by the fact that most of the valves that he was to access were in out-of-the-way places. He had to feel for each valve and, in near-darkness, use a mirror on his wrist to verify its status. Since physical work invariably involved applying a force, preparing the AMU was more difficult than expected. Almost every action created a 'push off' reaction that disturbed his position. In training, he had been able to set up the AMU rapidly, but the KC-135 aircraft had provided periods of weightlessness lasting for only 20 seconds, and performing the entire sequence of actions in space in real time proved to be more taxing. By the time that he had set all the valves he was hyperventilating, and in the darkness he did not notice the accumulation of mist on the inside of his visor until it was almost opaque. "This visor is sure fogging up," he told Stafford on the intercom. As he set about deploying the AMU's arms, which had first to be rotated down and then extended in a telescopic fashion, his heart rate rose to 180 beats per minute. It had been hoped that with his feet restrained he would have both hands free to work, but whenever he applied a force his boots slipped out of the stirrups and he drifted out of position, which obliged him to hold onto one of the handrails. As a remedy, he placed one boot in a stirrup and used his other to jam it in. He was able to relax only when he had buckled himself into the backpack.

"He's fogging real bad," Stafford reported when they rose over Carnarvon's horizon, "the oxygen is on 'High' to keep it to a minimum. He's got the attitude controller and manoeuvre controller arms out, but it was far more difficult than in the simulator."

"How're you doing, Gene?" Stafford enquired on the intercom.

"I'm really fogged up, Tom."

"Take a rest."

"I'll go ahead and make the electrical change-over."

After plugging the AMU's oxygen line into his ELSS, Cernan unplugged the spacecraft's umbilical. As this had provided power and communications in addition to oxygen, from this point he was to communicate by the UHF radio in the backpack, but the reception from his location behind the adapter was poor and Stafford could barely discern what he was saying in the static when Cernan transmitted. They waited in silence, with Cernan resting. The decision point would be at sunrise, which was when the attachment bolt was to be blown.

"Can you see out at all, Geno?" Stafford asked once they were in range of Hawaii. Cernan replied with a burst of static. "Can you read me, yes or no?" More static. "Your transmission is awfully garbled." It would be irresponsible to proceed. "I say it's a 'No-Go', because you can't see." He told Hawaii. As Cernan was able to hear Stafford, he switched back over to the spacecraft's umbilical in order to revert to the intercom.

"Houston agrees with the 'No-Go'," announced Gary Scott a few minutes later.

Cernan unstrapped from the AMU, scrambled 'up' over the rim of the adapter module, and groped his way along the handrail. Locating the hatch by feel, he swung it open. As soon as his feet appeared in the opening, Stafford drew them down onto the seat. Unfortunately, in doing so, Stafford nudged the Hasselblad with which he had taken pictures earlier in the EVA, and it floated out. Cernan saw it dimly through his semi-opaque visor and reached for it with one hand, but did not get a firm grip and it slipped away – which is why there are no pictures of Cernan's EVA to rival those of White's.

"We called it quits with the AMU," Stafford confirmed when the relay to Houston was established via California. "We had no choice."

"Okay," Armstrong acknowledged.

As Cernan rested in the sunlight, his visor gradually cleared up. Once he had retrieved the movie camera, Stafford suggested that he fetch the mirror from the docking bar, but when he moved forward to do so his respiration rate rose and his visor fogged again.

"I'm going to make the recommendation that we ingress before sunset," Stafford decided, deleting the airglow photography scheduled for just after sunset.

"We agree with that, Tom," Armstrong replied.

Cernan eased his feet beneath the edge of the instrument panel and began to force himself back into his seat, but it was a struggle because his suit had inflated in the vacuum and he had no flexibility at the knee. In fact, having been reinforced to protect against the AMU's efflux, his suit was even stiffer than the one that White had worn. As soon as Cernan thought he was inside he swung the hatch down, but it bounced off his helmet. Stafford cursed, and used the lanyard to draw the hatch down until the latch engaged, compressing Cernan in the process, at which point Cernan used the ratchet to lock the hatch. Stafford immediately repressurised the cabin to collapse his colleague's suit enough for him to relax his awkward posture. As soon as he could, Cernan removed his helmet. Seeing his companion's distress – his heart was racing at 180 beats per minute, and he was quite literally burning up – Stafford disregarded operating procedure for the nozzle that obliged the astronauts to take care not to spray water around in case it caused an electrical short circuit, and squirted a jet of cold water onto Cernan's face to cool him down.[4]

"Believe it or not," Stafford wryly observed a little later, "I think we've learned a lot."

"We certainly agree with that," Armstrong replied.

"I think it was a real fine exercise," Cernan agreed.

"You did good work, friend," interjected Dick Gordon, who was to make a spacewalk on Gemini 11 in a few months time.

"You don't know how much!" Cernan replied.

"Yes I do, I was watching [your heart rate]."

In light of the umbilical evaluation, it is clear if Cernan had been told to float over to the ATDA to try to release its shroud, he would not have been able to position himself to work effectively, and in floundering around may well have caught his suit on one of the shroud's many sharp fixtures, so it is lucky that the order was never given. However, if he had been able to don the AMU first, and used it to position

---

[4] During his exertion, Cernan had sweated off 10 pounds. This remained in his suit, making him chilly for the remainder of the mission, and on doffing his suit following recovery he poured water out from his boots.

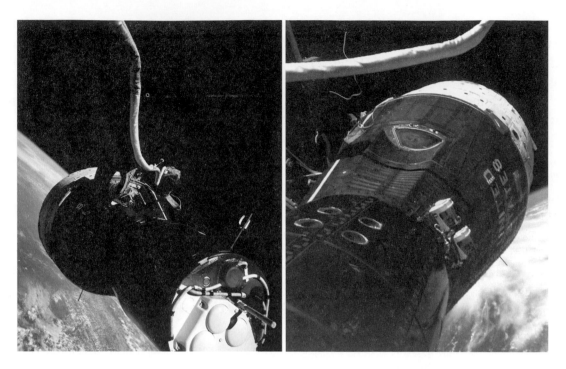

While outside Gene Cernan snapped a few pictures using a camera mounted on his chestpack. One (left) shows the rear-view mirror that he affixed to the docking bar and in the other Tom Stafford's ghostly face can be seen peering through his tiny window.

himself alongside the ATDA, it is possible that he would have been able to release the shroud to facilitate a docking, which would have been a powerful demonstration of the utility of a spacewalking astronaut.

In truth, Cernan had been the innocent victim of NASA's ignorance of how to undertake extravehicular activity. His stability aids and training had been inadequate. But Gemini's role as a pathfinder was to identify the limitations, and, in overcoming them, advance the state of the art. Thus, in determining that EVA was not as straightforward as White's experience had implied, Cernan had provided useful data. If Dave Scott had been able to go out on Gemini 8, he probably would have found donning the ESP backpack in the adapter was more difficult than expected, and it may well have been possible to provide Cernan with better hand rails and foot restraints. In the lottery of life, however, it was Cernan who learned this lesson. Despite extensive training, he later admitted that he "really had no idea how to work in slow motion." Major Ed Givens, the Air Force's project officer for the development of the AMU, suggested that having reached the point of waiting for Stafford to release him for independent flight, Cernan should have waited for his helmet to clear up and then at least have verified the basic functionality of the AMU.[5] In his debriefing, Cernan ventured that if the mobility issue could be overcome, when Aldrin flew on Gemini 12 he should be able to demonstrate the AMU to be "a flyable machine", so the project was not lost.

[5] Givens had been recruited by NASA earlier in 1966 as a member of the fifth group of astronauts. He died in an automobile accident on 7 June 1967.

Following Gemini 9, George Mueller appointed the Gemini Mission Review Board under the chairmanship of James Elms[6] to reassess the Program's goals, the manner in which they were being addressed, and the spate of problems that had compromised their achievement. As regards the fogging of Cernan's visor, because the succession of crews worked more or less independently to develop their flight procedures, Cernan was unaware that on Gemini 8 Scott was to have applied an anti-fog agent to the inside of the visor as part of the preparation for his EVA. This became standard procedure on future spacewalks. The overall conclusion from Cernan's ordeal, however, was that the training regime for spacewalking had to be improved.

Return

Shortly before retiring for the final sleep period, Stafford made Gemini 9's orbit elliptical by dropping a perigee 12 nautical miles in order to reduce the retrofire dispersions, and thereby improve the re-entry, and after an eventful and exhausting mission he steered the capsule to a remarkable splashdown within a mile of the *USS Wasp*. Because they were in the midst of the recovery force, which, at Stafford's request, had stationed itself right on the target, once the swimmers had fitted the flotation collar he opted to remain with his capsule and ride the crane to the carrier's deck.

Gemini 9 splashes down (left). Gene Cernan and Tom Stafford chat (right) with one of the swimmers as a helicopter hovers nearby. They opted to ride the capsule onto the deck of the *USS Wasp* before egressing.

The ATDA post-mortem

Once the close-up pictures of the ATDA's fouled shroud had been examined, the source of the error was identified. The Agena shrouds were manufactured by Douglas Aircraft and the GATVs were built by Lockheed, whose Cape technicians had been

---

[6] The Gemini Mission Review Board was formed by George Mueller on 23 June 1966 under the chairmanship of James Elms (Mueller's deputy) and comprised Charles Mathews (Gemini Program Manager), Edgar Cortright (Deputy Associate Director for Space Science and Applications) and Major-General Vincent Huston (of the Air Force's Eastern Test Range). Whilst its mandate was to assess planning for the final three Gemini missions, its focus on EVA planning resulted in it being informally referred to as the EVA Review Board.

trained to install the shrouds. The ATDA had been put together by McDonnell as a one-off and its technicians had installed the shroud themselves. Although they had followed what they believed to be the proper procedures, they had fitted the lanyards incorrectly and taped them into place when they ought not to have done so. The Lockheed supervisor had been called away and his verbal instructions for how to finish the job had been misunderstood.

On the positive side, the Air Force was impressed by the way in which Gemini 9 had performed the close-in fly-around of the ATDA to report on its condition, and hoped soon to have its own 'Blue Gemini' crews make similar reports on Soviet satellites.

Chapter 8

Orbital Operations
Gemini 10

A 'crazy' plan

While the Program's primary objective was to show that rendezvous was practical, it was apparent that doing so in lunar orbit would be more challenging than in Earth orbit, where the World-Wide Tracking Network could supply the data to enable Houston's large computers to calculate the sequence of manoeuvres. It was therefore decided to establish that a rendezvous could be made without ground support.

In September 1965, when the detailed planning for the Gemini 9 and 10 missions started, they were both to attempt the 'm=3' rendezvous. However, by the time that John Young and Michael Collins were assigned to Gemini 10 in January 1966, it had been decided to perform an exercise in optical navigation on the first revolution to assess the scope for using a sextant to measure the elevations of stars above the horizon as a preliminary to utilising a graphical technique to determine the orbit. The spacecraft's computer had been upgraded with a larger memory module to enable Collins to calculate the sequence of manoeuvres for an 'm=3' rendezvous – which would start with the second revolution and hence conclude on the fourth. Since Dick Carley, who had designed the upgrade, had promised that it would make Collins a modern Ferdinand Magellan,[1] Young had cheekily given his pilot–navigator this nickname.

On the assumption that an Agena from either Gemini 8 or Gemini 9 would be parked in high orbit following its primary mission, it was decided to demonstrate docked manoeuvres by having Gemini 10 ignite its Agena's Primary Propulsion System in order to use one Agena to reach another. Furthermore, once alongside the second target vehicle, Collins was to make a spacewalk to retrieve a micrometeoroid experiment. In light of the fact that the first orbital rendezvous had yet to be achieved when this plan was first outlined, the Gemini Program's rendezvous objectives were being addressed at a tremendous pace. Program Manager Charles Mathews described it as "the most ambitious of all the Gemini missions". On being shown the flight plan for a double rendezvous followed by a spacewalk to retrieve a package from the uncontrollable vehicle, Young said that the mission planners were "out of their minds". Since its battery would have expired, the old Agena's transponder would not

---

[1] The sixteenth century Portuguese explorer Ferdinand Magellan was the first to lead an expedition that would circumnavigate the world.

be operating and so the terminal phase would have to be performed without radar tracking. Nevertheless, on 18 May 1966, the day after an Atlas lost Gemini 9's Agena, Mathews confirmed that Gemini 10 would try docked manoeuvres. In response to continuing doubts by some managers about the reliability of the Agena's PPS, he said that the GATV had been *designed* to facilitate docked manoeuvres, the Program was rapidly drawing to an end, and it was time to "bite the bullet". After Gemini 9 confirmed that the terminal phase could indeed be done using optical tracking he confirmed that Gemini 10 would use its Agena to rendezvous with Gemini 8's Agena. In the event of its own Agena being lost, Gemini 10 would be directly inserted into an 87 by 208 nautical mile orbit, peaking marginally below the circular orbit of the old Agena, and perform an 'm=16' rendezvous so that Collins could conduct his spacewalk to retrieve the experiment package.

John Young (left) with Michael 'Magellan' Collins.

There were many issues to be considered, including when Gemini 10 ought to jettison its own Agena. The major manoeuvres to set up the phasing prior to establishing the co-elliptic 'catch up' orbit just below the passive target would clearly be performed by the Agena. But should it make the co-elliptic burn? Should it make the TPI burn? If so, could it perform the delicate braking manoeuvres of the terminal phase? If not, then would the Gemini be able to do it with a 7,000-pound Agena on its nose? And if not, could the Agena safely be jettisoned immediately after the TPI burn and travel alongside the Gemini spacecraft during the transfer? The only way to resolve these issues was to undertake comprehensive simulations and assess the various factors. Another issue was the timing of the second rendezvous. When should this be done? On the second day or on the third day? And how would this affect other activities? It was decided to start the second rendezvous utilising a phasing orbit with its apogee at 400 nautical miles – almost twice that of the target, which was decaying as a result of air drag. As the old Agena's transponder was inert, its orbit could be determined only by radars that could 'paint' its skin for a reflection. It was decided to establish the phasing for the rendezvous on the third day in order to give the World-Wide Tracking Network time to determine the phase relationships sufficiently accurately for Houston to calculate the optimum co-elliptic orbit to set up for the demanding optical terminal phase. The record-breaking high-apogee introduced the issue of exposing the crew to the trapped charged-particle radiation in the van Allen Belt. It was decided to orient the apogee to avoid the South Atlantic Anomaly, a zone off the coast of Brazil where the charged particles – mainly electrons and protons – trapped by the Earth's magnetic field dip down towards the ionosphere. The Trajectories and Orbits team therefore had a daunting task in drawing up a flight plan that satisfied all of these requirements.

EVA review

The first Apollo mission was expected early in 1967 so, if the final few Gemini missions were not to be cancelled, the Program had to maintain a fast pace with missions being flown at two-monthly intervals. However, as the training cycle was six months, it was difficult for crews to adjust their training to absorb the lessons learned by their immediate predecessors. The three-day Gemini 9 mission was in early June 1966 and the post-flight debriefing took all of ten days, so by the time that the lessons drawn from Cernan's extravehicular activity could be passed on to Collins his mission was only a month away.

The Gemini 10 patch.

Collins's speciality in the Astronaut Office was the development of the G4C suit, and EVA equipment (although, perversely, he had not been involved in the development of the Ventilation Control Module chestpack worn by Ed White) so he was delighted when he was assigned *two* EVAs. As some of his tasks had to be undertaken in darkness, he was to start his first outing at sunset and work standing in the hatch. On drawing up alongside Gemini 8's Agena, he was to go out to retrieve the micrometeoroid package that was to have been collected by Dave Scott[2]. Because Collins would have to operate on an unstabilised vehicle, this would be challenging, but he would be able to use a Hand-Held Manoeuvring Unit (IIIMU) to stabilise himself. In February 1966, it had been decided that even though Gemini 8 was to introduce a backpack with a tank of freon for the HHMU, the spacewalkers on Gemini 10 and Gemini 11 were to hook up a hose to a tank of nitrogen gas in the adapter module. The Gemini Mission Review Board formed after Gemini 9 concluded that Cernan's difficulties had been specific to attempting to work in the rear of the adapter. As Collins would be required only to plug his hose to the valve on the more accessible area on the surface of the adapter immediately aft of his hatch, the Board concluded that if improved handrails were fitted he ought to be able to achieve his objectives. In late June, NASA ordered a study into whether neutral buoyancy could enable weightlessness to be simulated with sufficient fidelity for astronauts to train for spacewalks. This had been suggested by a small aerospace research company in Baltimore in Maryland, which promptly hired

[2] Because Collins had been assigned the task of working on the inert Agena in January 1966, which was several months before Gemini 8 flew, if Scott had been able to attempt his spacewalk and retrieved this particular S-10 package, Collins would have retrieved the replacement that Scott was to have installed.

a school swimming pool in nearby McDonogh for the trials. Collins was too busy to rehearse, but other astronauts did and reported that he would have trouble with the spring-loaded nitrogen hose if it failed to engage on the first attempt, because he would require to use two hands to recock the connector, during which time he would drift out of position. This potential problem notwithstanding, at the pre-flight press conference he was up-beat about his prospects for retrieving the package from the Agena.

John Young and Michael Collins with their backups Al Bean and C.C. Williams at the pre-flight press conference.

An excellent start

In launching Gemini 10, the Cape's task was more complex than for previous rendezvous missions, because this time both the 10-Agena and Gemini 10 would have to be launched into the orbital plane of the 8-Agena.[3] Furthermore, during the four months that the 8-Agena had been in space, the plane of its orbit had precessed, which meant that the new launches would have to occur in the afternoon. Accordingly, on 18 July, Young and Collins were able to sleep until noon, which was a very civilised time to start a mission.

After the Atlas lifted off at 15:39 Cape time, GATV-5005 manoeuvred into the desired circular orbit. To be able to rendezvous with its own target, Gemini 10 would have to lift off within a 35-second window at 17:20, but to facilitate the dual rendezvous it would have to lift off precisely on time, which it did.

"There's no doubt about lift-off!" Collins exclaimed as they accelerated away. As they went supersonic, the aerodynamic pressure caused a severe vibration, but the buffeting soon diminished. The second stage ignited 'in the hole' and the blast ruptured the oxidiser tank at the top of the first stage. The resulting explosion made the staging appear spectacular to the tracking cameras, but by that time the upper stage had drawn clear. When the expanding cloud of gas momentarily swept over the spacecraft, Young realised that something was different to his ride on Gemini 3, but he was not overly concerned.

"Here comes the world!" Young advised as the second stage pitched over to start to build up its horizontal speed and gave the right-seater a great view downrange. "Isn't it beautiful."

"Goddamn, that really is," Collins agreed. And then, by way of an apology, he continued, "I promised 'no swearing', didn't I!?"

"You're right on the line," advised Gordon Cooper, the Houston CapCom,

---

[3] The Agena left by Gemini 8 was now being referred to as the '8-Agena' so as to distinguish it from the one to be launched specifically for Gemini 10, which was referred to as the '10-Agena'.

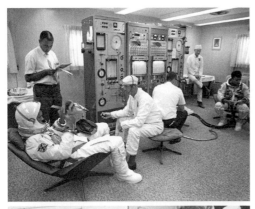

Preparations for Gemini 10: Michael Collins and John Young suit up in the trailer on Pad 16 (top); Young leads Collins from the trailer; and in the White Room Guenter Wendt (wearing glasses) presents Young with an oversized pair of pliers cut from foam.

meaning that their trajectory was following the nominal track on the plot on the main display at the front of the control room.

After 15 seconds of sightseeing, Young ordered a return to work; they had instruments to monitor. "Get your head back into the cockpit."

"'Go' for IVAR," Cooper advised at SECO. As the launch vehicle had left them with a shortfall of 27 feet per second, Young corrected this with the post-separation burn and the resulting 146 nautical mile apogee was off by only 1 mile. Their Agena was in a 160 nautical mile circular orbit and 900 miles ahead. The 8-Agena was 60 nautical miles above that, and about 500 miles behind.

"All personnel," Cooper announced, "we're going to debrief in the Ready Room." This had been intended for the ground team, but a miskeyed switch included Gemini 10.

"I hope they have a good debriefing," Young chuckled to Collins on the intercom.

"Sorry," Collins apologised dryly on the radio, "we're going to miss that debriefing!"

'Magellan'

As they flew out over the Atlantic, and the evening terminator loomed, Collins loaded the new memory module into the computer. "I'll be ready to start on the orbit determination as soon as I unstow this Kollsman sextant," he promised. "I see stars!" he noted when next he looked out. On taking a few sightings of the star Schedar, the brightest star in Cassiopeia, to test measuring the elevation of a star above the horizon, he found doing so to be more difficult than in the simulator. "John, I hate to be pessimistic, but I think this is going to be sort of bad." With no Moon to illuminate the Earth, the boundary between the dark limb of the planet and the black sky was ill-defined. "I'll do the best I can."

A multiple exposure showing the erector lowering prior to the launch of Gemini 10.

"That's all you can do," Young agreed.

"The real horizon isn't good enough, John," Collins pointed out after taking half a dozen measurements of the test star and finding an unacceptably wide spread in the results. As his eyes adapted to the darkness, however, he began to discern a line along the horizon. He had just over half an hour of orbital darkness in which to make a series of sightings of two stars, one directly ahead of the spacecraft and one behind it, and Young was to yaw around to face each star in turn. As Collins tried to measure Hamal, the brightest star in Aries, he realised that what he had presumed to be the limb was the top of the 'airglow' layer. Giving up on the Kollsman, he retrieved the Ilon sextant to find out whether he could discern the horizon any better with its tighter field of view. "It's very poorly defined," he complained. "It's just like *guessing*, darn it – I mean, it's not like, 'Oh, there's that line'." While Young yawed around to face back along their track, Collins retrieved the Kollsman. He followed Vega in Lyra down to what he thought was the horizon and timed its disappearance, only for it to momentarily reappear before finally fading away. "I can't find the ruddy horizon." He decided to switch to the backup star in Aquila. "We'll do old Altair."

"We're just about to come up on Carnarvon," Young pointed out.

"John, I'm really sorry, I muffed it," Collins apologised. He did not hold out much hope for the worth of his efforts. Nevertheless, each sighting had been entered into the computer to enable it to calculate the parameters of their orbit. "What's that?" he asked on looking out and seeing a multicoloured band delineating the horizon as Young yawed to face forward again. It was orbital sunrise. "That's amazing! Holy mackerel!" The rising Sun meant that he had run out of time. They had been so busy that they had not attempted to call Houston via any of the relay stations, but Bill Garvin in Carnarvon raised them for a brief update, and to give the '16-1' go-ahead. Collins had to use his orbit determination to calculate the phase adjust, plane change and co-elliptic manoeuvres. The plane change was the trickiest, as he had to calculate the node where their orbit intersected the target's orbit, and the fact that they were almost co-planar meant that he had first to calculate a very small angle of divergence and then compute the intersection point. The calculation used an iterative technique which (if given good data) would converge towards the solution, but the variance in his measurements was causing him difficulties. "This plane change isn't coming out right – it's all over the place."

"How're y'all doing up there?" enquired Ed Fendell casually as they rose above Hawaii's horizon for the first time.

"Just fine," Young replied, typically tersely.

Collins reported his phase adjust and co-elliptic manoeuvres, but was still working on the plane change. On receiving Houston's version of the sequence based on Carnarvon's tracking, Fendell began to read it up. "Wait a minute," Collins interrupted, "we're supposed to tell *you* first." Leaving Hawaii behind, Collins resumed his computations and his mood worsened with each passing minute. "Stupid, lousy chart. Let's start back here at the beginning." The iteration was *diverging*. "I'm sorry, John. It just isn't coming out. These darn residuals aren't coming down." Finally, he estimated the plane change to be 8 feet per second south. As they approached the coast of California he took a well-earned rest and soaked up the view. When Houston established contact via Guaymas in Mexico, he offered his plane change calculation without much confidence.

Crossing the United States towards the end of the first revolution, once again approaching orbital sunset, Collins was astonished to see "two beautiful stars right out my window."

"I don't have any," Young said.

"What the heck are they? Do you see that?"

"I bet those are the Agenas," Young mused.

"What?"

"Those have to be the Agenas; we're in the same orbital path as them." In fact, although their own Agena was ahead of them, the 8-Agena was far behind.

"Well, man," Collins exclaimed, "look at them. They're beautiful! What's that flashing on the right? Is that the Agena? See it flash? John, do you see what I see?"

"Yes."

"What the heck is that?"

"I dunno."

"Is this tape recorder still running?"

"Yes," Young confirmed. The tape recorded data from the spacecraft's systems as well as the crew's conversations, and it was periodically 'dumped' to ground stations.

Collins spoke for the tape, "They're at 12 o'clock. The time is 01:39:00. They're directly ahead. They appear to be co-altitude. There are two extremely bright stars – I'd say 1 degree apart. The one on top is slightly to the left of the one on the bottom. And about a degree below those two there is a flashing light, a smaller magnitude light, which is visible all the time, and which periodically flashes."

"There's no telling what they are," Young concluded.

"It's really weird," Collins continued for the tape. "When I first saw them, I could see the Earth lit by sunshine down in the lower right of the window and the spacecraft's nose is *still* in the light." The fact that the human eye is unable to resolve stars in space if there is a bright object within the field of view meant that the lights must have been *very* bright.

"Do you think we ought to report it to the ground?" Young asked. "No," he answered his own query, "let's not."

"See what they say," Collins prompted.

Young relented. "Houston, we have two bright objects up here in our orbital path. I don't think they're stars. They look like we're going right along with them."

"Roger," Cooper acknowledged. And since the crew had renewed contact, he relayed the decision on the rendezvous. "Your solution is 'No-Go'." Because their navigational solution differed significantly from that derived by radar tracking, Flight Director Glynn Lunney had decided to play safe.[4] "Where are the objects from you?"

---

[4] Collins's solution called for a 58 foot per second phasing burn at 02:19:52, an 8 foot per second plane change to the south at 02:53:25 and a 46 foot per second co-elliptic manoeuvre at 03:49:13. Houston's solution was a 56 foot per second phasing burn at 02:18:09, a 10 foot per second burn to the south at 02:30:49, and a 48 foot per second co-elliptic manoeuvre at 03:47:34. While there was general agreement on the in-plane manoeuvres, and both called for a southerly correction, the fact that Collins's plane change was 23 minutes late reflected his difficulty in calculating the nodal point.

Cooper asked, finally picking up on their call. "If you can give us a bearing, maybe we can track them down." He meant that the Air Force had a catalogue of every object in orbit.

"They just disappeared," Young replied. The objects, clearly satellites, had flown into the Earth's shadow. Moments later, as Gemini 10 flew beyond the Eastern Test Range, it did the same.

As they headed off across the Atlantic for the second time, Collins tried again with the sextant, this time with Fomalhaut in Piscis Austrinus and Arcturus in Boötes. "I think you're doing a tremendous job," Young encouraged.

"I'm doing a lousy job," Collins corrected[5].

"What the heck! If you can't see the stars, you can't see them," Young insisted. "I have been telling you that for 6 months." He had been skeptical of the experiment right from the start. It had merely confirmed the difficulty of sighting off the horizon. But they had not lost anything, because they still had Houston's solution for the rendezvous.

'We're running out of gas!'

The phase adjustment was performed over the Indian Ocean, at which time the 10-Agena was some 400 nautical miles ahead. The plane change was made a few minutes after skirting low over Carnarvon. With a PQI of 83 per cent they were in great shape. On making contact with Houston via Guaymas, Young noted that their radar had locked on at 220 nautical miles. Shortly before they set out over the Atlantic on their third revolution, Cooper gave them the revised data for the co-elliptic burn, and reported that the Agena was TDA-north to give the best view of its lights. Buzz Aldrin, 'spelling' Cooper, raised them through Tananarive to ask how the manoeuvre had gone.

"We burned 6 feet per second up and 48 forward," Young reported. The two-component velocity reflected the fact that they had pitched up at 8 degrees for the burn. They were now in an orbit 15 nautical miles below that of the Agena and catching up with their target, which was now only 100 nautical miles ahead. The rotation of the Earth meant that the *Coastal Sentry Quebec*, north of the Philippines, made its first contact of the mission. Fifteen minutes later, it was Hawaii. "We can see the target," Young reported. He had it centred in his reticle, and the pitch angle was rapidly increasing.

"Very good," Fendell replied. He gave Houston's latest data for the TPI burn, which was to be made several minutes after they dipped below his horizon. Young now faced a decision. Their own radar solution called for a forward component of 41 feet per second. Houston said 34 feet per second should be sufficient. Not unreasonably, he opted for the onboard solution. He spotted the Agena's acquisition

---

[5] On the third day, when the Moon was illuminating the dark side of the Earth, Collins resumed taking sightings with the sextant, and found that he could identify the horizon beneath the airglow. However, the tops of storms made it irregular, so measuring the elevations of stars was still not a straightforward process. In the post-flight debriefing he concluded that measuring the elevation of stars was "not viable" without moonlight. The verdict on the new memory module was also negative. As Young said, "I cannot understand how anybody could think that that is an operational thing that people ought to use in space flight."

lights soon after it entered the Earth's shadow. They flew most of the transfer far to the south of the United States. The *Rose Knot Victor* was stationed off the Atlantic coast of South America, and received their telemetry but the crew remained on the intercom circuit.

"I'm going to brake, babe," Young decided, on closing within 2 nautical miles of the target. Something had gone awry because they were not to have reached this point until shortly after sunrise. He would have to brake in darkness. To his alarm, as he did so, he noted that the PQI meter had fallen dramatically low. "We're running out of gas!"

"600 feet, *holding*," Collins reported. For some reason, they had drawn to a halt far from the target.

Young thrusted to resume moving towards the target, but they drew to a halt once again, this time at about 300 feet. "Whoa, whoa, whoa, you bum!" What was wrong? He nudged ahead. "How fast are we closing?"

"4 feet per second," Collins reported.

"Gosh darn, babe!" Young apologised. "I really missed this one. I don't have enough gas for you to do your docking practice." Once they had docked, Collins was to have undocked and redocked for experience.

"Ah, don't worry about it," Collins dismissed.

"I'd better get on in there," Young decided. To his consternation, the range began to open again, and he had to fight to close to 120 feet.

"Here comes the Sun," Collins noted as the horizon began to glow.

"Are you station-keeping yet?" Cooper prompted, on raising them via Tananarive. There was no response.

Young was preoccupied. As he had closed in on the Agena, it had passed out of the field of view of his window, and he was wary of a collision in the dark. "I can't see a darn thing here, babe!"

"I can see it," Collins assured. The Agena was to their right. Young started to yaw to the right. "You'll see it in a second."

"I've got it."

"What's your status?" Cooper persisted.

"Say again?" Young demanded.

"Are you there yet?"

"We're there," Young confirmed.

That was it for the moment. Young switched back to intercom. As the Manned Spacecraft Center's Public Affairs Officer pointed out in his commentary, "This has got to be one of the most untalkative spaceflights to date!" Furthermore, almost all of the reports had been by the commander. "We can only recall hearing from Mike Collins on two or three occasions."

"What's your range now?" Cooper asked after a few minutes of silence.

"About 40 feet," Young replied just before they dropped below Tananarive's horizon. As they flew on, he eased around to the TDA end of the Agena to enable Collins to examine the Status Display Panel.

The *Coastal Sentry Quebec* was to monitor the Agena during the docking. "You're looking real good," the CapCom, Gary Scott, reported upon acquiring the spacecraft's

The Gemini 10 crew were mighty glad finally to reach their Agena target vehicle.

signal. "How's it going?"

"We're just about ready to dock," Young confirmed.

"Does it look alright for docking?" Lunney asked Scott, asking about the Agena's status.

"Everything looks good here, Flight."

"Tell them."

"Everything looks okay for docking," Scott informed the astronauts.

"Let's cut in on this conversation," the Public Affairs Officer told his audience so that he could take a break, but there was only static. "That's almost predictable," he sighed, "when we expect this crew to talk, that's it! It's coming in very short spurts tonight."

"Do you see any docking yet?" Lunney prompted Scott.

"Negative."

"Docked yet?" Lunney prompted again a few minutes later.

"Negative."

"Did you get an OAMS quantity readout?"

"Will do," Scott promised. "Gemini 10. How's the docking?"

"We're still a couple of minutes away."

"Could you give us a PQI, please?"

"We're reading 36 per cent."

"Thirty-six!?" Scott echoed in amazement. Why so low on fuel? "Did you copy, Flight?"

"Yes, I sure did!"

At that point, the two spacecraft – still undocked – dipped below the ship's horizon. The infrequency of reports was frustrating for the flight controllers who were eager to know why the braking phase had used so much fuel. As Hawaii picked them up Fendell told Houston that the telemetry indicated that they had docked.

"You seem to have used a tremendous amount of fuel between *RKV* and *CSQ*,"

Fendell said when he raised the spacecraft. "Did something different than ordinary happen?"

"No. It just seemed like a tough break, I think," Young mused.

They had no time for troubleshooting, however, because before leaving Hawaii they had to yaw around 180 degrees to test the strength of the docking collar.

Revised plans

During the gap between Hawaii and the *Rose Knot Victor*, Tom Stafford, the astronaut with the most experience of rendezvous, discussed the extraordinary fuel consumption with the Flight Director, who opted for a contingency option in the flight plan. After using its PPS to put the docked combination into an orbit with a high apogee the Agena was to have been jettisoned, and after spending a night in this phasing orbit, Gemini 10 was to have descended to chase the 8-Agena, which by then would have drawn ahead. In order to save what was left of Gemini 10's OAMS propellant, Lunney decided to retain the 10-Agena in order to employ its engine for the 'drop down' manoeuvre.

The *Rose Knot Victor's* first item of business on acquiring Gemini 10 was to update its crew on the imminent PPS manoeuvre. Collins disabled his encoder to enable the ship to load the Agena's Velocity Meter with the required velocity change. This done, the ship's CapCom sought clarification of the propellant usage. Young recounted the two midcourse corrections after TPI and the manoeuvres during the final phase, in the hope that Houston would be able to figure out what had gone wrong. After TPI they had had 743 pounds of propellant. The terminal phase had been performed using the 'closed loop' solution. According to the nominal flight plan they should have had 680 pounds of fuel left, but had only 350 pounds.

"Did he spend a lot in the final braking?" Lunney enquired.

"I'll check with him," the CapCom said, and did so.

"I didn't think it was anywhere near that much," Young replied.

"Ask him if he felt his PQI pretty much followed the usage profile," Lunney prompted.

The ship attempted to relay this query, but communications faded out, and Houston was left wondering.

As would be determined following splashdown, when Gemini 10's inertial platform was initialised, it had been slightly misaligned with the result that the spacecraft was not flying in quite the plane that its platform indicated, and instead of reducing both the range and range-rate to zero simultaneously the 'closed loop' solution had brought the vehicle to a halt some 600 feet to the left of the target, and as the vehicles had pursued their slightly inclined orbits the range had varied as their trajectory traced a helical arc (or as Collins would later describe it, a "whifferdill") in to the target. They had seen such a situation in simulations when they had been out-of-plane, but this time their instruments had misled them into thinking that they were *in*-plane. The additional propellant had been consumed in eliminating this out-of-plane error. This was galling for Young, because in the astronaut corps efficient use of fuel was the primary indicator of having what author Tom Wolfe would later immortalise by the term "the Right Stuff". In the post-flight debriefing, Young regretted not splitting the difference between their own and Houston's TPI solutions. "That was my first error of judgement," he admitted. "I should have applied 37 feet per second forward, as near as I can figure, and let it go at that, and seen what developed." Since they had arrived early, they had had to brake in darkness. Furthermore, although he had observed that the target was drifting with respect to the stars, he had not tried to correct this because he was rehearsing the braking profile that he intended to use with the inert Agena. "I think it was an error to try to do more than one thing at a time. You've got to worry about one rendezvous at a time." The situation was complicated by the out-of-plane error – it was this that had caused the target to drift against the stars towards the end of the transfer. "With all that out-of-plane, I just couldn't hack it," he concluded.

The 'shunt engine'

By the time they reached the *Coastal Sentry Quebec*, the imperative was to make the final preparations for firing the Primary Propulsion System. "Everything looks 'Go' for the burn," advised Gary Scott.

Firing the Agena's engine would achieve another long-standing Program objective. Even if Gemini 6 had been able to rendezvous with its Agena, it would not have undertaken docked manoeuvres using either of the Agena's propulsion systems because, as demonstrated when the PPS exploded during the climb to orbit, it was regarded as an 'unproven' vehicle. Whilst Gemini 8 had been able to accomplish the historic docking, it had been obliged to undock and make an emergency descent, and in any case no docked manoeuvres had been planned. After Gemini 9's Agena was dumped into the Atlantic, the ATDA had been sent aloft but this was incapable of orbital manoeuvring. Gemini 10's successful docking therefore not only offered the first opportunity to attempt docked manoeuvres, it also had an ideal target in the shape of the 8-Agena. Utilising one Agena as a 'shunt engine' to chase another one would be a potent demonstration of what Wernher von Braun had long ago called 'orbital operations'. Nevertheless, as there was a significant risk that the PPS might misfire the tension in Houston – and indeed aboard Gemini 10 – was intense.

As the first burn would be prograde, the Agena was oriented with its PPS facing aft. At 16,000 pounds of thrust, the engine was expected to deliver a considerable 'kick' when it lit, and as the Gemini 10 would be flying 'blunt end forward' (BEF) it would be an 'eyeballs out' manoeuvre for the crew.

"Verify that the crew has put on their shoulder harnesses," Lunney prompted.

"Have you fastened all your restraint harnesses?" Gary Scott duly enquired of the crew.

"We're tightened down," Young assured. They had also stowed away loose items in order to prevent them becoming missiles within the confines of the cabin when the engine fired.

The burn was to be made while in contact with Hawaii so that the Agena's performance could be monitored. To initiate the manoeuvre, the low-thrust secondary propulsion system engines were to fire to provide ullage to settle the propellants in their tanks and then the PPS was to be ignited and fire for 14 seconds. As the Public

Affairs Officer warned, "Don't blink, or you'll miss it." But Young and Collins were not about to blink. If they detected any sign of an instability in the combustion they were to shut down the engine using a specially-installed 'Arm Stop' switch that was wired through the TDA directly to the Agena's engine controller because there would be no time for Collins to issue the command using his encoder.

"We're giving you a 'Go' for the burn," Fendell confirmed as they rose above Hawaii's horizon. "You're holding the attitude real fine," he added. The telemetry indicated that there was little thruster activity – the Agena's propellant was not sloshing and causing the docked combination to undergo excursions within the range of the 'dead band' setting of its Attitude Control System, so they were nicely set up for the manoeuvre. Then came the moment of truth: "PPS initiate!" Fendell reported for Houston's benefit, and then, after the programmed duration, ".... and shutdown!"

"That was really something!" Collins exclaimed enthusiastically. No one had ever seen a large rocket motor fire in space and, facing backwards, they had had a magnificent view.

The plume from the Agena's Primary Propulsion System was so dazzling that it prevented Michael Collins from observing the warning light on the Status Display Panel.

"Pretty wild, huh?" asked Fendell.

"When that baby lights," Collins continued, "there's no doubt about it!"

"We've got a really spectacular tail-off going right now," Young pointed out. After the engine had shut down, it continued to vent propellant. Since the burn had occurred over the sunset terminator, the efflux was back-lit by sunlight, and a spectacular sight. Collins would later write that the Agena shut down "coughing and belching", and they were treated to "a glorious Fourth of July spectacle". A "golden halo" surrounded the Agena. "In its centre, like a Roman candle, the engine was spitting out residual chunks of fuel and oxidiser, some tiny fireflies, others as large as luminous basketballs." As the display had lasted for 30 seconds, they had been able to unstrap a Hasselblad and take a few photographs.

"Don't turn the recorder off," Fendell advised. Prior to the burn, Collins had activated the Agena's recorder to document the engine telemetry, and Fendell wanted to ensure that it ran long enough to catch the engine's protracted tail-off. At Lunney's request, he asked whether the Agena had been stable during the burn.

"We had a little yaw," Collins admitted, "but got it right back."

The Agena's 3-degree excursion in yaw had introduced a slight out-of-plane element to the burn, but this would be able to be corrected later. The Flight Dynamics Officer reported that the new orbit was 160 by 412 nautical miles – considering that the apogee objective was 410 miles, this was excellent.

"I saw sparks flying, and noise – rattling," Young observed to Collins after they had left Hawaii behind.

"I almost shut it down; I almost did," Collins pointed out.

"No, you didn't!?"

"I almost did," Collins insisted. "If you'd said, I'd have shut it down. Really!"

"I was too busy hanging on the wall," Young chuckled.

"I came pretty close to shutting it down. I kid you not."

"Son of a gun!" Young exclaimed.

In the post-flight debriefing, Young recalled: "The first sensation I got was that there was a pop, then there was a big explosion and a clang, and we were thrown forward in our seats. Fire and sparks started coming out of the back end of that rascal. The light was something fierce." Collins later said that at first he was "disappointed to see only a string of snowballs shooting out of the back of the Agena in a widening cone." Then, as he was about to suggest that it had failed to ignite: "Wham! The whole sky turned orange-white" and it "kicked like a mule." He was to have monitored the engine's performance via the Status Display Panel on the TDA, but the plume was so bright that he had not been able to see the lights. In fact, the SDP instruments were originally to have been inside the Gemini spacecraft and electrically interfaced via the TDA but when it was realised that there was no space for them in the cabin they had been put on the TDA – where they proved to be impossible to read when it really mattered. Collins opined that it was unrealistic to expect the Gemini crew to shut off the PPS if a problem developed. "If that thing starts to go out of control, I think you would be hard pressed to get it shut down before it broke up." Young concurred: "I think it would be pretty close."

As they climbed towards their unprecedented high apogee, attention turned towards the dosimeters. The readings were amazingly low. "We're wondering if your dosimeters are still snubbed," Aldrin asked via Tananarive. Collins confirmed that the readings were barely 10 per cent of the expected level. The alignment of the apogee had been chosen to avoid where the inner van Allen belt dips down several hundred miles over the South Atlantic, so as to minimise the exposure to the crew.

Since the PPS burn had been made at sunset, they made the climb in darkness and did not get their first 'high' view until apogee, at sunrise. Unfortunately, since the Agena's Attitude Control System maintained a horizontal configuration, and since it blocked their view of the Earth, they saw only a sliver of the horizon out to each side. "I was tempted to turn the ACS off and just let it drift," Young recalled later, "and just see what we could see."

"We're standing by for the crew status report," Gary Scott called, as they flew into range of the *Coastal Sentry Quebec*.

"Crew status is 'Go'!" Young replied emphatically.

"Have you had anything to eat today?"

"Just a couple of meals, and some goodies we carried in our pockets." The issue of the astronauts carrying foodstuffs aboard in their pockets had been relaxed since the furore that had erupted after Young had handed Gus Grissom a corned beef sandwich during Gemini 3.

The phasing orbit's high apogee was designed to enable the 8-Agena, travelling slowly in its 216 nautical mile circular orbit, to catch up with and pull ahead of the docked combination. As Young and Collins retired for their first night's sleep, they did not yet know whether they would be able to complete the second rendezvous. It had originally been intended to undock from the 10-Agena at this point to carry out a series of redocking tests under different lighting conditions, but the excessive use of fuel during the final phase of the first rendezvous had prompted the retention of the Agena to enable it to perform the manoeuvres to establish the second rendezvous. The first of these 'extra' burns was the height adjustment at an elapsed time of 20 hours and 21 minutes. This retrograde burn was to lower the docked combination's apogee to 209 nautical miles, 7 miles below the 8-Agena's orbit, and several hundred miles in trail of it. This would be circularised one-and-a-half revolutions later by the co-elliptic burn. The Agena would finally be discarded, and Gemini 10's OAMS would make the TPI burn as planned. When Young and Collins were awakened, they were delighted to be advised that the Trajectories and Orbits team had refined the manoeuvres for the second rendezvous. "Hats off to those guys," Young praised afterwards, "they did a good job."

"You've got a 'Go' for your PPS burn," the Canary CapCom called as Gemini 10 passed overhead in the run up to the first of these manoeuvres, the height adjustment.

"Did you check his VM?" asked Flight Director Cliff Charlesworth, referring to the value in the Agena's Velocity Meter that specified the desired change in velocity. On being set, the meter was repeatedly checked to ensure that its content did not become corrupted.

"Affirmative."

In this high orbit, there was almost a seamless handover between Canary Island and Kano in Nigeria.

"It was a good burn," Young reported.

"Roger," replied C.C. Williams. He and Al Bean were the backup crew, and they were alternating on the CapCom console in Houston.

"It may be only 1-g, but it's the biggest 1-g we ever saw!" Young emphasised. Because the ignition of the PPS was almost instantaneous, firing the engine was a case of 0 g to 1 g in the blink of an eye.

As they headed out across the Atlantic on the 14th revolution, Canary set the Agena's VM for the final burn, which would last just 4 seconds. Loading the meter had been left until the last possible minute so as to incorporate the latest radar tracking of the intermediate orbit produced by the first burn in the calculation of the next manoeuvre.

"Have you got all your hot scoop?" Charlesworth prompted Bill Garvin as Gemini 10 approached Carnarvon, climbing towards apogee.

"If you mean the backup data for the circularisation burn, yes." Then he called

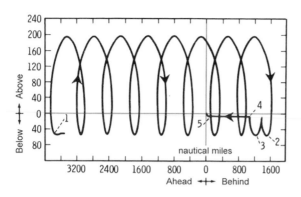

How Gemini 10 rendezvoused with the Agena left by Gemini 8.

1. Phasing manoeuvre at 07:38:34 (GET)
2. Height adjust at 20:21:02
3. Plane change at 21:50:49
4. Co-elliptic manoeuvre at 22:37:54
5. TPI at 47:30:41

Gemini 10: "We're looking at you here, and everything looks very good."

"Say they're 'Go' for the burn," the Flight Director insisted.

"You're 'Go' for the PPS burn."

"Roger," acknowledged Collins.

"You got all the info you need?"

"I believe so, thank you."

As the burn was made shortly after dipping below Carnarvon's horizon, they gave their report via Canton Island. They were now in a 206 by 209 nautical mile circular orbit. No sooner had Young read down the residuals than they left Canton behind. It had been decided to retain the 10-Agena a little longer in order to avoid using OAMS propellant for attitude control, and so it was still on the nose when Collins made his first EVA.

Stand-up

When Gemini 10 rose over Hawaii's horizon a few minutes later, Fendell asked Houston whether he should ask the crew to configure the Agena to enable its tape of burn data to be downloaded, but Charlesworth refused. "No, I don't want to bother them. They're in the middle of EVA preparations. Let's hold off, and get it later." The Flight Director then told the various stations on the ground network to closely monitor the spacecraft's telemetry for any indication that (as had happened on previous missions) an astronaut inadvertently hit a switch while handling the bulky EVA apparatus. As Young and Collins worked methodically through their checklist, they made only occasional comments to the ground. This prolonged silence was frustrating to the PAO in Houston, because the world's press, eager to find out if Collins would have as exhausting an experience as had Cernan, were listening in.

"John, this is Deke." Slayton rarely came on the circuit, but he was the boss, so when he did so it was to make a point. "You guys are doing a commendable job of maintaining radio silence. As soon as the French stop shooting at you," he began, making a topical reference to the atmospheric nuclear test that the French had just conducted over one of their Polynesian islands, fortunately while Gemini 10 was around the opposite side of the planet, "why don't you do a little more talking from here on?"

"What would you like us to talk about?" Young asked innocently.

"Well, anything that seems appropriate – like EVA."

"We've completed our final preparations," Collins reported. "We're just standing by for sunset." And that was it! However, after a brief silence, he called: "If you've got some camera experts down there, the 16-mm movie camera on the right-hand side is broken. It just started making strange noises and it got to the point where it would tick as if it were turning, but the little ratchet inside that advances film isn't moving. Have you got any suggestions for repair? I don't mean right now, but prior to the EVA tomorrow."

"We'll check into it," Williams promised.

When Collins fell silent, Williams decided to report the terrestrial news. "Gemini 10, you might be interested to know that the Astros dropped four straight to the Mets in New York."

Collins didn't take the bait, however. "It'll be nice to get this door opened and see what the world looks like." Their view ahead and down was restricted because the Agena preferred to fly 'straight and level', and so he was eager to see the Earth from a higher orbit than any of their predecessors, knowing that it would be particularly impressive in the wide view through his faceplate.

"I'm ready to go, John," Collins announced once the Sun had set. In fact, he was jumping the gun by a few minutes. After starting over Canary Island, their track would take them over Ascension Island, Kano, Tananarive, Carnarvon, Canton Island and Hawaii before starting the run across North America. If all went to plan, he would close the hatch after sunrise, over the western United States.

"Okay, babe," Young decided, "Go to it."

After the difficulties experienced with the hatch on earlier flights, McDonnell had made its mechanism operate at a lower force level, so Collins was able to open it easily, but in the process he imparted forces which set the spacecraft oscillating slightly, which prompted the Agena to fire its thrusters to maintain itself 'straight and level' with the TDA forward. With the hatch wide, Collins stood on his seat and dragged out the trash bag. "It will be nice to get rid of all this stuff," he pointed out, because the cabin was cramped enough without storing miscellaneous rubbish. Young was to use the surviving movie camera to film its departure in the twilight. "I'll fling it where you can see it. How about that! Look out of your window."

"I'll never get to it in time," Young complained.

"Okay, John, I'm going to set up the bracket, here." Collins had to mount the bracket for the 70-mm camera for the ultraviolet astrophotography experiment.[6] Because his eyes had yet to adjust to the darkness, he had to do this by feel, which was difficult because there was little tactile feedback through the gloves of his spacesuit. "This getting out at night is a loser, because I can't really see how my hoses and whatnot are doing." Canary Island reported its acquisition, and said that it would be standing by. Apart from basic communications checks, the stations were to remain off the air until the astronauts called on them. Believing that the camera was set up, Collins fired off an exposure to test it, but the camera became detached. (It was later determined that a screw on the bracket had worked loose and was inhibiting the camera from fully engaging the bracket.) As his eyes adapted to the darkness, Collins found that he could see what he was doing. After re-attaching the camera, he started

---

[6] Karl Henize, the principal investigator for the ultraviolet astrophotography experiment, became an astronaut in August 1967.

taking pictures of specific parts of the far southern sky. With the Agena TDA-forward, he was pointing the camera across Young's hatch. An automatic timer had been provided to operate the shutter, but Collins had attached it early and forgotten to stow the camera before the final brief firing of the Agena's engine, and when the camera hit the footwell the timer had snapped off. Collins was therefore firing the shutter manually, with Young timing the exposure. When NASA's Office of Space Science and Applications announced that Gemini astronauts would conduct experiments during spacewalks, astronomers suggested ultraviolet stellar photography. As ultraviolet radiation is absorbed by the atmosphere, this part of the spectrum can be studied only by sending instruments into space. The photographs could not be taken from inside the spacecraft because the windows had been coated to filter out ultraviolet light to protect the crew from the Sun. This was the first time that this task had been assigned, and it was timely because the first Orbiting Astronomical Observatory, which had been launched on 8 April 1966 on an Atlas-Agena, had been crippled by an electrical fault in its power supply during its activation procedure.

"Can you recognise any of those things out there, babe?" Young asked.

"Yes." Collins had his outer visor raised, and now that his eyes had adapted he could see the stars. Although from his vantage point above the atmosphere the stars shone brilliantly without twinkling, he was mildly disappointed not to see fainter than about fifth magnitude. "It's really nice out here, John," he observed. His suit was not chilling down in darkness. In fact, he was quite warm. "Can I talk to the ground?" Young switched him from the intercom over to the radio. "Everything is going pretty well up here, and we're a little ahead on time," he called 'in the blind', but there was no reply. "Can you hear them talking?"

"No."

"Maybe they're not getting us."

"Apparently not."

After a few more pictures, Collins tried again. "Houston, this is Gemini 10. How do you read?"

"I'm not reading a soul," Young pointed out.

As he worked through the list of star fields, Collins continued his radio commentary for the tape recorder. At one point he noted that even though the cabin was depressurised, loose items acted as if in an airflow – the tie-down on Young's helmet, for example, was "trying to sneak out the open hatch."

"I wonder if they can hear you yet," Young said.

"Who are we supposed to be reading?"

"On the ground?"

"Yes."

"I dunno," Young admitted. "One of those remote sites."

As they reached the midway point through the 20-picture sequence, the radio seemed to come alive as if someone was trying to call them. "Houston," Collins called, "so far, body positioning has been absolutely no problem; as a matter of fact, I sort of have to struggle to move up and down in the hatch. The suit, when it's pressurised, just fills the available space, so that there are plenty of points of suspension." On this occasion, stability was not really an issue and he was able to work efficiently. "It takes a bit of getting used to," he continued, "but I'm getting to the point, now, where I'm starting to enjoy it out here." In fact, his heart rate was a mere 100 beats per minute. In his relaxed state, he was not overly taxing his suit's environmental system. As a precaution – in light of Cernan's experience – Collins had wiped the inner surface of his faceplate with an anti-fogging compound before donning his helmet. "I don't think they're reading us," he mused, "or they'd be yakking it up."

"Houston," Young called, "have you been reading our conversation?"

"Yes," came the terse reply.

"I'm looking forward to the Sun coming up," said Collins. Not only would this be a great sight in its own right, he would be able to appreciate the view of the Earth. In the darkness, he could dimly recognise familiar outlines. Finishing the assigned star fields ahead of time, he decided to keep going until he ran out of film – after all, he had nothing else to do until after sunrise. Then he saw something low on the horizon shining brightly. "What the heck is that out there? Doggone, it's bright!" He had a burst of inspiration: "You don't suppose it's the old Agena, do you?" It would be the object of their second rendezvous. "I'll bet it is. We're charging along behind and below it." It had caught the sunlight ahead of them. "Houston, if you read, the Sun is just beginning to come up. To the east is an extremely bright object. It's approximately 8 degrees above the horizon. Could it be the Gemini 8 Agena?"

"Do you notice it moving relative to the stars?" Carnarvon asked.

"I haven't noticed any movement," Collins after watching it, "and unfortunately the stars are disappearing because the Sun's starting to come up."[7]

Finished the astrophotography, Collins dismounted the camera from its bracket and gave it to Young to have the film changed for the next experiment, in which he was to take a series of pictures of a colour chart for Houston's photo lab employing different exposures in order to calibrate the film – once they knew how the film responded, they would be able to render future space pictures in 'true colours'.

"Got your sun visor down, babe?" Young prompted.

"No, not yet."

When Young handed Collins the colour chart and the telescoping rod that was to hold it in front of the camera, the chart broke free as he tried to fasten it to the rod, but he managed to grab it and complete the assembly. But then he had to halt. "I've got a problem here, John," Collins called on the intercom.

"What's the matter, babe?"

"As soon as the Sun came up, my eyes started watering. I'm not sure whether it is this compound on the inner surface of the visor, or what it is, but my eyes are really watering like crazy – to the point where it's real difficult to keep them open to see what I'm doing."

"Don't look at the Sun," Young prompted.

"I don't think it's a function of the Sun. I've got my eyes closed. I can tuck my

---

[7] UFO fans cite this as a 'confirmed' flying saucer sighting. A more mundane alternative to it being the Agena left by Gemini 8 is that Collins had spotted his own trash bag close by catching the first rays from the Sun.

head in like a turtle and get down inside the suit, so my head is in the shadow."

"Mine have been watering too, babe," Young admitted. "The whole time."

"Is that right?" Collins exclaimed. Young had suffered in silence so that Collins would not to have to curtail his astrophotography and close the hatch. Young switched Collins over to the radio and he reiterated his story, but there was no response – they were several minutes beyond Carnarvon's range.

"Come on back in," Young decided. "Let's close the hatch."

In contrast to his predecessors, Collins was able to regain his seat readily. He had devised his own procedure. After jamming his toes under the edge of the instrument panel, he shuffled until his knees followed suit, then he used his knees as a fulcrum and forced his bottom down onto the seat. He swung the hatch down without difficulty and as soon as it was locked the cabin was repressurised .

"I still can't see anything," Young pointed out.

"Just close your eyes, John. It goes away after a while. Don't sweat a thing, buddy."

"It must be something in the oxygen circuit."

Collins thought he recognised the distinctive smell from a ground test a few years earlier. "I'll bet you it's something in the lithium hydroxide." The Environmental Control System recycled air through canisters of lithium hydroxide to remove carbon dioxide. It was a well tested system, so why should it release an irritant at this time?

At this point, Houston announced its acquisition via the Canton Island relay station.

"The hatch is closed and we're repressurising," Collins replied. "Something in the ECS has caused our eyes to water, to the point where we couldn't see. It actually smells. I don't know whether it's lithium hydroxide, or what. At first, I thought it might be the coating on the inside of my visor because that is the only thing I could think of which is new, but now I'm fairly sure it's not that."

"It happened to both you and John, is that correct?"

"That's right," Young confirmed. He began to explain that they had terminated the EVA at sunrise, but they had already left Canton behind.

"When did you first notice the problem, John," Fendell asked, when Hawaii picked them up a few minutes later.

"Just about sunrise," Young replied. However, he had a confession. "I was crying a little through the night. I didn't say anything about it because I figured I was just being a sissy." At that time, Collins had been dictating to the tape just how comfortable he was, standing in the hatch. On hearing that the eyes of both crewmen were "red and slightly swollen", the Flight Surgeon in Houston recommended that they use a damp cloth to soothe their eyes. Houston and Hawaii went into conference over the ground network's loop, checking the telemetry to establish the precise configuration of the ECS – occasionally asking the crew to verify switch settings. The immediate recommendation was to close their visors and switch the oxygen flow to 'High' because this would bypass the lithium hydroxide, but by this time the problem was abating. Whilst Collins's early ingress had meant cancelling the daylight photography, the fact that he had finished the astrophotography meant it was a successful spacewalk. However, the second EVA would be at risk of cancellation until the cause of the eye irritation was identified and resolved, as it would be folly to risk having both men rendered blind when Collins was alongside the inert Agena and Young had to be able to see to preclude a collision. In Houston, the ECS engineers set about investigating possible causes of the problem.

The EVA had been squeezed in between the circularisation manoeuvre and a refinement of the catch-up rate of the rendezvous with the 8-Agena. As this involved only a small increase in velocity, the Secondary Propulsion System was to be used.

"SPS ignition, Flight." Fendell reported from his telemetry.

"Roger."

"We have cut-off!"

"Thank you."

"It appears to me that the burn was just a little bit long," Fendell advised Houston. It was about 10 rather than 7.7 feet per second. Whilst this would change the timing of the terminal phase somewhat, it would be manageable.

As Gemini 10 left Hawaii, the Public Affairs Officer resumed his commentary. "Several people have remarked, here, on the extraordinary versatility of the Agena. It has proved itself beyond the fondest hope during this period of nearly 24 hours now, both in terms of small burns like that we just saw over Hawaii and the large burns that we saw last night and earlier today. It is truly a remarkable vehicle, and its performance has been precisely as advertised." In fact, the 'shunt engine' manoeuvres using the 10-Agena convincingly 'ticked the box' for the docked manoeuvres. Although Gemini 10 had made up for all the previous frustrations in this regard, the mission's greatest challenges were still to come.

Second rendezvous

"Good morning!" cheerfully called Gary Scott from the *Coastal Sentry Quebec* when he awakened Young and Collins after an 8-hour sleep to begin their third 'day' in space. This done, he signed off because they had barely poked above his horizon, and then because the spacecraft would not pass over the ship again for some time, Scott retired to sleep so as to be fresh to assist with the de-orbit sequence that would lead to a splashdown the following day. The immediate task in space was to correct the slight out-of-plane error imparted by the yaw excursion during the big engine's first firing.

"We checked the VM, and all systems are 'Go' for your SPS burn," Carnarvon advised.

With 10 minutes to go, Young realised that they were pointing the wrong way for a burn to the north! He had earlier told Collins to use his encoder to 'gyrocompass' the Agena TDA-*south*, and it was only when he recognised a northern constellation that he noticed the error, so they "rather swiftly" yawed around TDA-north. The burn was made over the western Pacific, but the ground track did not offer many communications passes and Houston had no word until they clipped the Eastern Test Range and Grand Turk Island got a brief relay. "How was your last burn, John?" asked C.C. Williams.

"Oh, it was okay," Young replied matter-of-factly.

The final phase adjustment was executed about 20 minutes later, and was monitored by Canary Island. "We've got SPS-ready," the CapCom advised Houston. "Attitudes are holding good. Start of burn! ... End of burn!"

"Roger," the Flight Director acknowledged.

"I think this thing's still overshooting a little," Young opined in reporting the residuals. "I guess *we* probably shut it down."

"We didn't see a VM cut-off," the CapCom added for Houston's benefit. It seemed that Collins had terminated the five-second burn manually by hitting the shut down switch.

Gemini 10 was now 246 nautical miles behind the 8-Agena, and slowly catching up. Thus far, although it was being conducted rather higher than usual, this rendezvous had been fairly routine. The catch-up phase was being pursued slowly, to enable the WWTN radars to establish the lighting requirements for the terminal phase – which would be the really demanding part of the experiment.

"I guess Mike's looking forward to getting the elephant off, isn't he?" Williams said on the next pass over the Eastern Test Range

"He sure is," Young agreed. With the conical docking collar of the Agena blocking their view, he said having the Agena on their nose was "like a railroad engineer driving down the road with a big freight train, and all you can see is the freight train."

"How's it feel to be rid of that 'freight train'?" Williams enquired via Kano shortly after the Agena had been jettisoned.

"It was a mighty *good* train!" Young pointed out.

"It sure was, John," Williams agreed. "For your information, at the time you separated from your Agena, the 8-Agena was 138 nautical miles."

"What's our delta-H?" Young asked.

"It's 7 miles."

"So we're right on it, aren't we?"

"We're going to sweeten everything up with these two tweak burns, John."

"That'll be fine," Young said, appreciative of the ground team's effort to create a perfect run in to TPI. "That sure was a good Agena."

"Let's go find the other one, now," Williams urged.

"You bet!"

As the 8-Agena's battery had long-since expired, its radar transponder was unavailable, so this rendezvous would be particularly tricky. Some ground radars were powerful enough to track the vehicle by 'painting' its skin, but the spacecraft's radar could not do this. Houston intended to steer Gemini 10 precisely to the TPI point, but thereafter the crew would have to rely on visual sightings, which meant making the transfer in daylight. Furthermore, since their target had no lights, the braking phase would have to be finished prior to sunset, lest the two vehicles collide in the ensuing period of darkness. If all went well, Young would then hold the old Agena in the beam of his docking light while Collins prepared to venture out at sunrise. In light of the diminished OAMS propellant, there was little margin for error. Even so, Williams assured them that the Flight Dynamics Officer was "very optimistic" about their chances.

"All systems are 'Go'," assured Garvin in Carnarvon, prior to reading up the data for the two forthcoming OAMS burns.

"How about the pitch and yaw?" Lunney prompted in the belief that Garvin had omitted some information.[8]

"S-O-P says 'don't do that'."

"Give it to him," the Flight Director insisted, "I think you're wrong."

"Roger," Garvin acceded. "Gemini 10, do you want the pitch and yaw for that?"

"No," Young replied, "that's all right."

"He doesn't want them, Flight."

"Yeah, I copied," Lunney granted. "I'm checking the S-O-P, though!"

"Page 4-9," Garvin offered helpfully.

"You win," Lunney conceded. "I'm wrong."

"Roger, Flight."

"We have the 8-Agena in sight," Young announced when Canton Island picked them up. "We've been watching it for about 5 minutes." He was astonished to be able to resolve it as a cylinder. The fluctuations in brightness indicated that it was slowly tumbling, which was bad news because Collins would not be able to retrieve the experiment package if the vehicle was unstable.

"We'll standby and watch your burn," Fendell called from Hawaii, several minutes later. The first tweak was imminent. "Burning, Flight," he told Houston, and when he saw in the telemetry that Young was firing thrusters to zero the IVIs, he finished his report with, "He's finishing up."

"With any luck at all, you'll hit TPI within 4 seconds of nominal," Williams announced when Guaymas acquired the spacecraft. This was precision flying, indeed. As Gemini 10 had flown over the Pacific, the trajectory analysts had been double-checking Young's report that he had the target in sight. "The Agena you saw, was it the 8-Agena? Or the 10-Agena?"

"Is the 10-Agena ahead of us?" Young asked, as the truth dawned on him.

"That's affirmative, about 3 miles."

"That's what we're looking at," Young agreed. He had spied the wrong vehicle! This was why he had been able to discern its shape. Obviously, their Agena had been disturbed as they had separated from it, and it was slowly tumbling. "That's too bad, C.C., I really thought we were seeing something."

"95 miles is pretty long range," Williams teased.

"You have to have real good eyesight for that," Young agreed. In the debriefing after the mission, Collins found it hilarious that on 'spotting' something at a range of 95 nautical miles, "instead of thinking 'That's the 10-Agena, you idiot!', I thought

---

[8] This bit of banter on the ground network is included here because it is amusing, and also because it provides insight into how the ground support system operated.

'Boy, I've got good eyes'."

"We burned that one right down," Young reported via the Corpus Christi relay station in Texas, referring to the second tweak. These final set up manoeuvres had fixed the differential height between the two orbits at a constant 7 nautical miles with an unmeasurable ellipticity – perfect for a rendezvous.

"I've some mission rules for you," Williams called, as Gemini 10 flew down the Eastern Test Range on its 29th revolution. "Your fuel cut-off to stop the rendezvous – to stop what you're doing at that point – is 7 per cent. If you arrive with greater than 10 per cent, you're okay to go ahead with station-keeping." This reflected the fact that OAMS propellant would be required to drop a perigee from the higher-than-usual orbit prior to the de-orbit sequence. And since fuel would inevitably be used in station-keeping during Collins's spacewalk, they were authorised to proceed with this only if they had at least 10 per cent when they arrived alongside the 8-Agena. Having completed the set up, Gemini 10 had 31 per cent of its OAMS propellant, which was barely sufficient for the TPI burn and the subsequent braking.

"I'd like to know the position of the 10-Agena when we're at TPI, relative to us," Young prompted on making contact with Canary Island. He was determined not to put his reticle on the wrong vehicle.

"Standby," the CapCom stalled. The focus of everyone's attention was the 8-Agena, the 10-Agena was history.

"We're running down the Agena 10's position now," Lunney told Canary, "but it's going to take a couple of minutes."

"LOS," Canary reported when Gemini 10 flew out of range.

"We'll get it to them," the Flight Director promised.

When the ranges and elevations had been calculated, Williams relayed the information via Tananarive. At sunrise, the 8-Agena would be at a range of about 29 miles nautical, and at an elevation of 13.5 degrees. At TPI, 23 minutes after sunrise, the range would have halved and the angle would be 33 degrees – this was about 5 degrees higher than the canonical case due to the fact that the height difference was only 7 nautical miles. There was no need for Young to concern himself about sighting on the wrong vehicle because by TPI the elevation of the 10-Agena would be *minus* 19 degrees.

As had Stafford on Gemini 9, Young rolled his vehicle inverted to put the Sun behind its nose so as to enable him to view the target in daylight without his reticle being flooded with light.

"Don't lose that rascal," Collins urged.

As Gemini 10 approached Hawaii, Lunney called Fendell: "You're within a couple of minutes of the TPI burn, Ed."

"Gemini telemetry is solid," Fendell confirmed.

"Just give him a 'standing by'." He was not to disturb the crew. If they required his help, they would call him.

"We're standing by," Fendell called up to the spacecraft, "watching for your burn."

"Roger, we have him in sight," Young replied.

"TPI in 40 seconds," Collins noted on the intercom.

"See if you can mark the burn time for us," Flight prompted Fendell.

"Start of the burn, Flight. He's quite a bit early."

"Yes," Lunney agreed. "He's about 14 seconds early."

The predicted timing notwithstanding, Young had made the transfer when the pitch angle was right.

"He's ceased burning," Fendell reported. A moment later, seeing a resumption of thruster activity, he added, "It looks like he's going to burn some more."

"Does it look like residual burnout, Ed?"

"That's about it."

"Do you have some residuals for me?" Fendell asked Gemini 10.

"Not at the moment," Collins came back tersely. They were busy. The burn had not gone very well – Young was unsure how long he had thrusted, and Collins, believing that they had picked up a 28 rather than a 25 feet per second increment, was working out how this would affect their first midcourse correction.

"Okay, standing by, I've plenty of time," Fendell pointed out.

"Never mind their residuals, Ed," Lunney decided. "Let's wait and see what he does." It was up to the astronauts now; the ground sites were mere observers.

"Okay, Flight. He's doing a big burn here using his down-firing thrusters. We're getting a lot of OAMS right and left activity." With Young continuing to consume fuel, the spacecraft slipped below Hawaii's horizon.

Because the target's transponder was inert, the radar could not supply range or range-rate. Instead, Collins was to use the sight of his narrow-field sextant to estimate the apparent size of the target and then calculate its range. He was not optimistic of his chances. When Collins said they were 2 nautical miles out and closing at 50 feet per second, Young started to brake. Ninety seconds later, Collins opined that they were right on track, at slightly less than a mile. The next measurement indicated about half a mile. "I'm starting to believe this data, John!" Working to keep the target centred in his reticle, Young barely acknowledged the succession of updates. Soon after Collins estimated that they had a quarter of a mile to go, the range-rate tailed off. It appeared as if, as on the rendezvous with their own Agena, they were drawing to a halt short of their target. "I'd thrust *towards* him, John!" Collins urged. Recalling Houston's order that they break off if they ran low on fuel, Collins checked the PQI gauge. "You're *fat* on gas, John." Things got better. "Don't brake any yet," Collins warned. As had Schirra when viewing Gemini 7 in daylight, Young found that as they closed in he was able to judge his range fairly accurately by the Agena's size in the reticle.

"See anything of the 8-Agena around?" Al Bean asked eagerly when Guaymas picked up Gemini 10.

"Yeah," Young replied. "We're about – I guess – seven or eight hundred feet out."

"Fantastic, John!"

"Yes, I don't believe it myself."

"We do," Bean assured. "What's your PQI?"

"You go ahead and fly, babe," Collins told Young. "I'll talk to those guys." Then to the ground he replied: "A little over 20 per cent, Al."

"Good show!!"

The Public Affairs Officer was impressed too. It was "far and away the most economical rendezvous transfer manoeuvre made." Charles Mathews, the Gemini Program Manager, and Bill Schneider, the Mission Director in Washington, had come into the MOCR to monitor the final phase of this demanding mission.

But Gemini 10 wasn't there yet, and as they set off down the Eastern Test Range Gary Coen, the Guidance, Navigation & Control Officer, warned that Young was "really hitting it" in closing the last few hundred feet.

"We're station-keeping," Collins announced a few minutes later. "John, can you read the PQI?"

"12 per cent."

"It's more than that," Collins insisted.

Young moved his head to view the gauge square on. "About 15 per cent."

On being assigned the 'shunt engine' mission to chase an inert Agena, Young had said that the planners were "out of their minds", but they had done it! Given that the cut-off for the station-keeping exercise had been set at 10 per cent, they were in great shape for the EVA.

"What's the attitude of the Agena?" Bean asked.

"Engine down," Collins replied. The Agena had been flying 'straight and level' when Houston had last heard from it, but over the months the Earth's gravity had tugged on its heavy propulsion system, which was serving as an 'anchor' to hold it vertical with the TDA on top.

"It's pretty well stabilised then?"

"Solid as a rock!" Collins said happily. The fact that the Agena was stable was excellent news because it suggested that retrieving the experiment package should be straightforward.

Before the vehicles flew into the Earth's shadow, Young manoeuvred to face the Agena, and maintained it in the beam of his docking light while Collins prepared to go out at sunrise.

Working on the inert Agena

Tests soon determined that the eye irritation during the first EVA had arisen because (as specified in the flight plan) two compressor fans had been active at the same time, and that the problem would not recur if only one fan was used. Lunney gave the good news to Garvin, the Carnarvon CapCom, who passed it on to Gemini 10: "You have a 'Go' for the rest of the station-keeping."

"How about the EVA?" Collins asked.

"That's what we mean."

"I'm glad you said that," Young observed, "because Mike's going outside right now."

"Good luck, Mike," said Garvin.

Collins rotated in the open hatch to face aft, reached back to the adapter and depressed a button to release the pop-out handrail that he was to use whilst hooking up the nitrogen hose for his HHMU.

"Watch that thruster there, babe," Collins warned, referring to the manoeuvre thruster on the retro section immediately aft of the spar between the two hatches, which would spew hot gas at him if it were to be fired to translate downward.

"I'm going to have to translate down in a second," Young warned, because they were so close to the Agena.

Collins swung around on the handrail until he was clear of the thruster. "Go ahead."

"Just one little pulse," Young confirmed. This done, Collins made his way to the adapter to retrieve the micrometeoroid package that he would have retrieved on his first outing if that had not been curtailed. "If I don't translate soon," Young noted, "we're going to run into that buzzard."

"Wait!" Collins manoeuvred clear again. "Okay, go ahead."

"There we go."

Collins rotated to face the hatch and poked the package, which was a slab about 6 inches wide and 1 foot in length, down through the hatch. Although Young accepted it and stuffed it between his knees as an expedient, he lost track of it, and it floated out at some point and was lost.

As they were beyond Carnarvon and Hawaii was some minutes away, Collins dictated a commentary to the tape recorder. "Everything is going well. I'm taking a lot more time to do each task than I'd anticipated. Right now, I'm trying to get the nitrogen line connected. I've retrieved the package. We've spent a lot of the time simply holding the spacecraft relative to the Agena."

"Can we back out a little," Young suggested. During these preliminaries, Collins' actions were disturbing Gemini 10, causing Young to repeatedly adjust his attitude to keep the Agena in the narrow field of view of his window. He wanted to withdraw until Collins was ready to make the crossing. In fact, it would have been better to have completed the preliminaries prior to moving up to the Agena.

"You want to go down?"

"No. I want to back out."

"Okay, back out," Collins allowed. The forward-firing thrusters would not threaten him. Young opened the range, with Collins holding onto the handrail for the ride. "Don't go *down* until I tell you," Collins reminded. He set off along the handrail, hand over hand to the flap in the skin of the adapter, which he opened to gain access to the valve of the nitrogen tank. The hose connector was designed so that when he pushed it onto the valve, a cocked sleeve would automatically slide down and engage. In weightless training, he had been able to hold himself stable on the rail using one hand and attach the hose using the other. However, the connector did not engage properly and he had to let go of the rail to reset the sleeve, and when he swung his arm to reset the sleeve his body twisted, his legs bumped the spacecraft, he bounced off and began to tumble. He had to keep grabbing the rail to stabilise himself as he recocked the connector.

"Boy, Mike!" Young exclaimed as the spacecraft rocked. "Take it easy back there."

On his second attempt, Collins successfully engaged the valve. "Okay, I'm hooked into the nitrogen." To make it more manageable, the hose was bound to his umbilical several feet along its length. Before moving to the Agena, he was to loop the spare section of the hose and push it under the rail to keep it tight against the adapter in order to preclude it straying across the aperture of the nearby thruster, but the rail had only partially deployed and he was unable to snub the hose. "I'm coming back to the cockpit area for a second," he told Young.

"I have to translate up, okay?"

"Go ahead and translate up, that's alright." Collins held on while Young made the move, and then returned to the hatch to stuff the loop of hose down into the hatch. "You're going to have to snub that down some place," he told Young. "Can you do that? I'll watch the Agena while you take care of the nitrogen line."

"How was it, to get it hooked up, Mike?" asked Young as he drew in the loop of hose and jammed it between the seats.

"It wasn't hard; it's all body positioning – like they said." He had been warned that if the hose failed to engage the valve on the first attempt, he would have difficulties.

By now, Gemini 10 was about 20 feet from the Agena. "I'm going to put you right next to it," Young decided. He intended to position Collins to enable him to reach over and retrieve the package. Collins held on as Young oriented Gemini 10 parallel to the Agena and started to close in, but at about 8 feet Young could barely see the vehicle in his tiny window. "If I go in any closer to it, I won't be able to see it, or you."

"Can you get a little bit closer," Collins asked. "I'll give you directions."

"Okay," Young acceded after a few seconds of pondering the situation.

"Go forward just a bit."

"Forward," Young confirmed.

"Better stop right there, John. Aft – translate aft. Okay. You're in a good position." The Agena was 6 feet away with the TDA directly opposite the open hatch.[9] All Young could see of the Agena was the far end of the engine unit.

"I'm going to leap for her," Collins announced.

"Take it easy, babe."

Pushing off gently in order as not to build up too great a speed, Collins floated across to the Agena. His aim was accurate, and he managed to grab hold of the conical docking collar and halt his motion without destabilising himself. He then 'walked' his way hand over hand around the collar near to where the package was located on the vehicle's cylindrical surface. Unfortunately, on reaching with one hand for the package, his body twisted, he lost his grip of the smooth metal of the cone and he cartwheeled away from the Agena.[10]

"Where are you, Mike?"

I'm back behind the cockpit, so don't fire any thrusters." He had spun around, and was now looking down past the vehicles at the Earth below. Unfortunately, his contact with the Agena had set it slowly tumbling. As he drifted by the open hatch, he grabbed it to stabilise himself. Then he resumed his jump-off position and drew his hose clear

---

[9] All directions here are relative to the Gemini spacecraft. In fact, since the Agena was "engine down", so was the nose of the Gemini.

[10] Given that it had always been intended that an astronaut would retrieve this experiment, it is unfortunate that no handrail was fitted to the Agena.

of the problematic thruster. "Okay, John. Let's try it one more time."

"How's that?" Young asked, having relocated about 10 feet from the now unstable Agena to give himself a wider field of view.

"That's fine," Collins accepted. "I'm going to try the gun, this time. I'm not having much luck without it." However, on squirting the HHMU to set off, his boot clipped the hatch and the disturbance sent him on a trajectory that did not lead to the Agena, and when he squirted again to try to correct, the impulse was not directed through his centre of mass and induced a slow rotation. As he drifted past the Agena he reached for the docking collar and managed to grasp a wire harness between the cone and the body of the vehicle – fortunately, the electrical system was inert. With such a firm grip, he was able to halt his fly-by, stabilise himself, and then work his way around the collar once again.

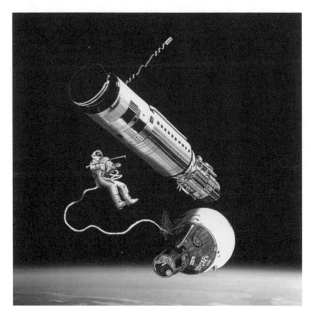

A depiction by artist Ed Hengeveld of Michael Collins manoeuvring to retrieve a package from the 8-Agena.

"There's a lot of garbage on there," Young warned, referring to a metal whisker used to discharge an accumulated electrostatic charge, which had come loose and was poking out from the collar.

This time, Collins was moving very slowly so as not to build up momentum and slip off, but he was disorientated. "Am I going the right way?"

"Are you coming back here?" Young asked, puzzled.

"No, I'm looking for the package."

"It's around the other side," Young told him. "See that you don't get tangled up in that fouled thing."

"Yes, I see it coming." Collins gingerly swung himself beyond the discharger. Young watched to ensure that the umbilical did not snag on it. Once in place, Collins kept a grip of the wire harness with one hand and reached to retrieve the package with the other. He had first to depress two buttons to release the spring-loaded fairing, which opened easily, prior to lifting the package out of its slot, in the process snapping the thin wire which had been installed to prevent it from flying out when the fairing was released.

"Have you got it?"

"Yes!"

"Come on back," Young urged. "Get out of all that garbage." By now Collins's activities had seriously destabilised the Agena, and the risk of his umbilical snagging increased. Young had always believed that working on an inert vehicle would be risky, and he was reluctant to push their luck.

"Don't worry, here I come." Collins tugged on his umbilical to set himself moving back slowly towards the Gemini.

"Do you want me to turn the spacecraft around to meet you?" Young offered.

"Don't do a thing," Collins countered. "And don't fire any thrusters if you can help it."

"You don't see the Agena anywhere, do you?" Young asked urgently. Having lost sight of the unstable vehicle, and being unable to manoeuvre, he was concerned about a collision. Even a glancing blow could easily damage his spacecraft. The worst case scenario was that Collins would be squashed between the two vehicles.

"Yes, I see it." They were well clear of the Agena.

As Collins didn't seem concerned, Young relaxed.

Meanwhile, the Flight Director alerted Fendell in Hawaii: "He's coming at you, Ed."

"How're you doing up there?" Fendell called once communications had been established.

"Old Mike went over there and picked up the package alright," Young replied. At that moment, Collins thrust the package in through the open hatch for him to stow.

"Did you put the new one on?" Fendell asked.

"No – and we're not going to," Young replied. "A bit of the TDA has come off, and Mike sort of got tangled up in it, so we'd better not fool with it any more." The experiment was to have been replaced, but the Agena was now so unstable, and there was very little chance of anyone ever retrieving the replacement. The original was to have been retrieved by Dave Scott within a matter of hours of the Agena being launched, so the fact that it had been exposed to the space environment for three months ought to satisfy the experimenters.

Fendell told Houston that his telemetry was poor. With Gemini 10 in an unusual attitude, its antennas were not favourably oriented.

"John," Collins called urgently, "the Agena is right behind you, to your left, so translate to your right."

Young was uncertain of Collins's motivation. "Do you want me to go back to her?"

"Translate to your right!"

Young manoeuvred away from the Agena.

"Tell them not to spend any more fuel in trying to stay with the Agena," Lunney ordered Fendell when the telemetry firmed up and it became clear how much fuel had been consumed in station-keeping between Carnarvon and Hawaii – it was down to 8 per cent.

Collins began to prepare for his next assignment, which was to evaluate the efficiency of the HHMU: Ed White had shown that it worked; Dave Scott was to have donned a backpack with a tank of gas to subject it to a full 'qualification trial', but the Gemini 8 mission had been cut short and Collins had inherited this step-by-step assessment.[11] As had been intended for Gemini 8, the test was to involve dragging the spacewalker along by his umbilical, but Lunney decided that they did not have the propellant and he ordered them to cancel the test.

"Get back in," Young told Collins.

After disconnecting the nitrogen hose, Collins took a short rest standing in the hatch.

On establishing contact via the relay site in at Point Arguello in California, Houston asked for a PQI reading.

"Get serious!" Young dismissed. With a cabin full of umbilical, he could not even *see* the gauge. Houston made intermittent calls over the next 15 minutes as Gemini 10 flew across the United States to no response. "We can't believe it," Young suddenly announced without preamble, "we'd turned off the radio and we've just found out!"

"That's what we suspected," noted Bean. He read a list of other switches which the flight controllers had noticed them inadvertently hit while wrestling to stow the EVA apparatus.

As an explanation, Young pointed out that having the 45-foot umbilical loose in the cabin "made the snake house at the zoo look like a Sunday school picnic." An hour after closing the hatch, Collins swung it open again briefly to jettison the EVA apparatus and other trash.

"That was a great job, today," Houston congratulated them later.

"It was a tremendous thrill," Young agreed. "It was really incredible. I don't believe part of it myself!"

Several hours later, Gemini 10 prepared to break out of the high circular orbit. The plan called for a 100 foot per second burn, but the OAMS would peter out when the regulator fell below its minimum operating pressure, and the engineers predicted that they would get only 75 feet per second.

"We burned 100, and we have still got fuel," Young reported via Tananarive, and then he shamefully pointed out that while the hatch had been open to jettison the trash the flight plan – their most important document – had slipped out as well. "So you might keep us abreast of what's going on."

The Public Affairs Officer reported the effect of this news: "I think everybody got the best laugh out of the mission from that."

All round success

Gemini 10 splashed down 4 miles from the *USS Guadalcanal* in the South Pacific east of Australia, drawing to an end a remarkable mission of orbital operations, a high phasing orbit, a double rendezvous and two successful spacewalks.

In contrast to Cernan's EVA, which was widely considered to have been a failure because he did not test the AMU, Collins's excursion was hailed as a success because he achieved his primary objective of retrieving the package from the Agena.

[11] Once the HHMU had been proven, a future mission, most likely one associated with the development of an orbital laboratory, would be able to address the vexed issue of donning a backpack for working independently of the spacecraft. By 1966, NASA was planning an Apollo Applications Program, and one of the ideas was to convert a spent S-IV stage into an enclosure in which to test EVA procedures in safety.

John Young is winched up into the recovery helicopter (left).
A delighted Young and Collins on the deck of the *USS Guadalcanal*.

Similarly, although his first EVA had been curtailed, he had done the astrophotography without difficulty. The cancellation of the HHMU evaluation was no loss, because it could be passed on to the next mission in line. Collins's apparent success went a long way towards alleviating some of the doubts which had set in concerning extravehicular activity. He was to have mounted his 16-mm movie camera outside to record his encounter with the 8-Agena but it had malfunctioned, and whilst it had been decided that Young should use the other one and shoot through his window whenever an opportunity presented itself, being in so close to the Agena he had been preoccupied avoiding a collision, and so Collins's adventure went completely undocumented. Nevertheless, Gemini 10 marked a milestone by being the first of the multiple-objective missions to achieve *all* of its primary tasks, and as such it set the standard for the remainder of the Program.

Chapter 9

Setting Records
Gemini 11

Standing out

Deke Slayton assigned Gemini 11 to Pete Conrad and Dick Gordon on 21 March 1966 as a straightforward rotation after backing up Gemini 8. The primary objective was a first-orbit rendezvous, a demanding assignment that called for an intensive series of manoeuvres leading to interception barely an hour after launch. The intention was to assess how an Apollo lunar module might abort its descent to the Moon and make a rapid return to its mothership.[1]

Previously, the IVAR manoeuvre had corrected only the fore/aft velocity error, but in an attempt at an 'm=1' rendezvous this would have to be combined with the up/down and out-of-plane corrections which on an 'm=3' rendezvous had been made during the second or third revolutions. After using a stick to sketch out the options on the beach at the Cape with Neil Armstrong, who had rotated to back him up, Conrad chose to make the plane change in two parts. The out-of-plane IVAR component would make the inclination of Gemini 11's orbital plane parallel that of the target vehicle, and a second burn, about half an hour later when 90 degrees from the insertion point, would make it coplanar. Of course, the out-of-plane error would be trivial if the lift-off was on schedule and the Titan's upper stage steered true, but the launch window was so narrow – only 2 seconds – that some managers, including Chris Kraft, doubted that Gemini 11 would make it. Whilst it would be feasible to launch within the next 30 seconds and make an 'm=3' rendezvous so as to pursue the other objectives, the fact that the Apollo 1 mission had slipped to February 1967 meant that there was now plenty of time in which to mount the final Gemini missions, and it had been decided that if Gemini 11 missed the window for 'm=1' the launch would be rescheduled to the next such opportunity.

In the expectation that Gemini 10 would demonstrate the Agena's ability to undertake docked manoeuvres, the issue became how Gemini 11 should exploit this new capability? Although the 10-Agena would still be available as a passive target, as the 8-Agena had been for Gemini 10's dual rendezvous, Conrad was reluctant simply to repeat his predecessors' feat – he wanted to do something that would make his mission 'stand out' from the others.

---

[1] At that time, it was expected that a lunar module would fly an 'm=3' rendezvous following a nominal lift-off, but this was later improved to 'm=2', and eventually to 'm=1'.

Dick Gordon and Pete Conrad pose
for their official crew portrait.

In mid-1965, while in training for Gemini 5, Conrad had heard about a plan called 'Large Earth Orbit' in which a Gemini spacecraft would be dispatched on a loop around the Moon if it appeared that the Soviets would beat Apollo into cislunar space. Because the Agena lacked the power for the 'translunar injection' burn, the idea was for the spacecraft to dock with the more powerful Centaur upper stage, previously placed into orbit by an Atlas. However, the development of the cryogenic engines for the Centaur had fallen behind schedule, the vehicle had not been adapted to carry a docking adapter, and the Gemini spacecraft's heat shield was not designed for the greater thermal stress of such a high-speed re-entry.[2] At an early stage, James Webb had ruled out contingency planning for a 'circumlunar' Gemini mission because it would merely draw resources away from that programme – which was at that time in the throes of a funding crisis. Denied the Moon, Conrad sought something else that was likely to capture the imagination. One idea was to use the Agena to rendezvous with another satellite. In 1965 NASA had launched three satellites on Saturn I rockets. Once in orbit, the 'boilerplate' Apollo capsule on the S-IV stage unfolded a pair of long thin panels which contained sensors to report strikes by micrometeoroids. In light of their shape, these satellites had been named Pegasus after the mythical 'winged horse'. Conrad suggested rendezvousing with one simply because the wings – spanning 100 feet – would be a spectacular sight, but this was ruled out. Finally, he settled for establishing an altitude record by flying a "very high orbit". Gemini 10's high apogee had been designed to establish the phasing for the second rendezvous, but lacking a second target Conrad sought an *independent* rationale for flying a high apogee. At that time, NASA's meteorological satellites flew at 675 nautical miles. Although they transmitted crude monochrome television imagery to Earth, the US Weather Bureau was considering installing a colour camera on its next generation of satellites. Conrad convinced the meteorologists that if Gemini 11 used its Agena to raise its apogee to that altitude, it would be able to snap pictures on film for comparison with the television from the satellites to help in assessing the merit of colour in meteorological work. When Gemini 10 found the radiation to be less intense than predicted at 412 nautical miles, Conrad was given permission to go higher – although with the safety proviso limiting Gemini 11 to only two high passes.

Conrad also accepted another task which – if he managed to pull it off – promised to be spectacular. In late 1965 it had been suggested that a spacewalker on one of the

[2] In fact, one of the Program's goals was to develop 'controlled re-entry' techniques which would enable an Apollo spacecraft returning from the Moon to slow itself by a 'skip' manoeuvre prior to making its re-entry.

The Gemini 11 patch features the use of the Agena to attain a record apogee (of 750 nautical miles).

later flights hook up a tether to the Agena in order to facilitate two engineering experiments. In one case, the Gemini spacecraft would position itself directly above the Agena to test the ability of the tether to maintain the vehicles in a stable configuration in the 'gravity gradient'. This had no direct relevance to Apollo, but might find a role in the Apollo Applications Program, where it would permit two vehicles to station-keep without the need to expend propellant. This might be useful in assembling a large structure in space. In the second test, inspired by the popular perception of a space station as a 'wheel in space', the tethered system was to be set rotating around its centre of mass in order to assess the scope for creating 'artificial gravity'.

All in all, Gemini 11 promised to be a mission for the history books, which is precisely what Conrad had in mind.

A double-exposure depicting Gemini 11 leaving Pad 19
an hour and a half after an Atlas lifted off with its Agena target vehicle.

Gemini 11 does it in one

On 11 August NASA announced that Gemini 11 would be launched on 9

September, but the count was scrubbed even before Conrad and Gordon had left the Crew Quarters when a 'pinhole' leak was detected in the oxidiser tank of the first stage of the Titan. The following day, as they were in transit to the pad, they were recalled because a fault had been found in the Atlas's autopilot. After breakfasting with Al Shepard on Monday, 12 September, Conrad and Gordon drove to Pad 16 to suit up and, on arriving at Pad 19 at 07:25 Cape time, were briefed by their backups, Armstrong and Bill Anders, who had configured the spacecraft. In the integrity check, it was found that Conrad's hatch leaked, but this was fixed in short order. The Atlas launched on time at 08:05, and GATV-5006 manoeuvred into the required parking orbit. Gemini 11's count was held at T–3 minutes so that the launch window could be refined using Carnarvon's radar tracking, and the Titan's guidance system updated. Gemini 11 lifted off at 09:42:26.5 – half a second later than planned, but still good enough for an 'm=1' rendezvous.

Another enterprising photographer snapped Gemini 11's liftoff with a mock-up of a Saturn V on Pad 39, known as 'Moonport, USA', in background.

"Go, you muthah!" Conrad urged happily on the intercom. "Roll program," he reported more calmly on the radio to Houston.

"Roger, roll," confirmed John Young, who was serving as CapCom for the launch phase.

"Man, we're on our way!" Conrad enthused as the Titan accelerated through 80,000 feet.

"Isn't that something!" Gordon agreed

"Feels great," Conrad assured. He had ridden a Titan before, and was satisfied with this one's performance. "The son-of-a-gun is just about alright."

"You're 'Go' for staging," Young advised.

"Staging, and engine ignition," Conrad reported.

The Flight Dynamics Officer assured the Flight Director that the Titan was "right down the middle". The second stage's guidance system was steering for the desired orbital insertion point.

"Standby for 'point-8'," Young advised as they attained 80 per cent of the speed required to achieve orbit. "Mark!"

"SECO!" Conrad called a few moments later.

After a 2-second burst of the aft OAMS had moved them clear of the spent stage, Conrad pitched the spacecraft's nose down to the horizon to burn the IVAR's forward component. The Titan had left them with a 39 foot per second shortfall, so he immediately corrected this to refine the apogee – from which they were to initiate the terminal phase of the rendezvous. The 1 foot per second out-of-plane error was so minor that he decided to ignore it, because it would be able to be incorporated into the 90-degree correction. When Gordon calculated the 15 foot per second downward radial velocity correction, Conrad burned it. As Gemini 11 left the Eastern Test Range behind, it was off to an excellent start.

"Gee, the time is really going by," Conrad observed as they worked methodically through their post-insertion checklist.

"It sure the heck is!"

"You're in an 87 by 151 nautical mile orbit," Young relayed through Ascension Island.

"Sounds perfect," Gordon agreed.

On spotting the Agena visually, Conrad tried the radar but got only intermittent returns. After yawing around for the 3 foot per second northward plane change, he resumed SEF and gave the radar another look, and this time received a solid lock-on. "Be advised, we're inside of 50 miles," Conrad said when they rose above the horizon of Tananarive on the island of Madagascar, "and we have the Agena in sight."

Young gave the backup data for the TPI burn. The plan was to initiate the terminal phase when Gemini 11 was 18.9 nautical miles behind and 8.6 miles below the target, which Young advised would be just after apogee, at an elapsed time of 50 minutes. The altitude differential would be somewhat less than the canonical case, so the transfer would subtend a central angle of only 120 degrees.

As they approached the sunset terminator, Conrad kept his spacecraft pitched up to the Agena. As soon as his eyes adapted to the darkness and he spotted the target's acquisition lights, he centred it in his reticle. After verifying that he could hold it steady using the radar needles, he resumed optical tracking. Meanwhile, Gordon monitored the computer. It used the target's increasing elevation and decreasing range to calculate the 'closed loop' solution for TPI comprising of 140 feet per second forward, 27 down and 5 left. Houston's figures were 139.6, 17, and 6.6. Gordon's own calculations indicated 140, 22 and 4. With such excellent agreement, Conrad decided to accept the computer's values, dipped the nose 8 degrees below the horizon and nudged it over to the left so as to burn all three components in one long firing. "We're burning, right now," he told Bill Garvin when Carnarvon picked them up.

Twelve minutes later, having left Carnarvon behind, Conrad made the first mid-transfer correction, once again accepting the computer's solution. But then he saw that even though the target was fixed in his reticle, the radar needles had started to stray, as if the strength of the radar signal was inconstant. Gordon used his encoder to switch the Agena to its spiral antenna in the hope of boosting the signal, but the MAP light did not illuminate to confirm that the command had been received and enacted.[3] He commanded it to revert to the dipole antenna, again without receiving confirmation. When the computer gave its solution, it was clearly rubbish. However, because Conrad could see that the target was not drifting against the stars he was confident that they were on track and he omitted the second correction.

"Tell them you're standing by," the Flight Director told Ed Fendell, once Hawaii acquired the spacecraft's telemetry. "We missed him at Canton."

"Gemini 11, Hawaii. Standing by."

"We're at 15,000 feet, and closing at roughly 50 feet per second," Conrad replied. "I have the Agena's running lights." Pitched up at 90 degrees, they were directly below the Agena.

"42 feet per second, Pete; a mile and a half," Gordon advised.

"I believe I'll go ahead and brake a little bit," Conrad decided.

"One mile; 6,000 feet."

"He's bright!!" Conrad exclaimed, as they flew into daylight. The sunlight reflecting off the Agena was so dazzling that it took a while for his eyes to adapt to the darkness of the cabin to read his instruments, so he felt for his sunglasses.

"Here, use mine," Gordon offered.

"Never mind," Conrad decided, "I'd never get them inside the helmet anyway."

"You're out in front of him, just a little bit," Gordon noted. They were now pitched up beyond vertical.

"Let me know when we hit half a mile."

"Yessir!" Five seconds later, Gordon called, "You're half a mile, right now."

Conrad slowed to 20 feet per second.

"Quarter of a mile; 1,500 feet."

"That's about the brightest thing I've ever seen!" Conrad complained as he slowed to 15 feet per second.

As they approached 500 feet, Gordon ceased making calls because, like his predecessors, Conrad was able to judge range and range-rate visually by the size of the Agena in his reticle. At this point, they left Hawaii behind. Although the braking was going well, Gemini 10 had demonstrated there was scope for a sting in the tail, so the tension in the Mission Operations Control Room soared. As Gemini 11 closed in, Gordon commanded the Agena to switch off its acquisition lights; once again there was no MAP confirmation, but the lights did go out. At 50 feet Conrad moved around

---

[3] This was the Message Acceptance Pulse (MAP) lamp on the Gemini spacecraft's instrument panel.

to give Gordon a view of the Status Display Panel. Despite the absence of the confirmation signal, the Agena seemed to be okay, which was good because if there was the slightest doubt about its status the managers were likely to cancel the docked manoeuvres.

"Gemini 11, this is Houston through California," Young announced in order to let them know that they were back in communication. "Standing by."

"We're station-keeping, looking at the TDA," Gordon reported.

"Outstanding!" Young congratulated. This news prompted a round of applause from the flight controllers.

"John, tell Mr Kraft – would he believe 'm=1'?"

"He believes it!" Young assured.

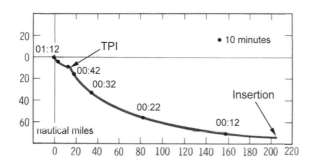

How Gemini 11 achieved the demanding first-orbit rendezvous, initiating the terminal phase transfer from near its initial apogee.

"Are we 'Go' for docking?" Conrad asked.

Firstly, Houston wanted to study the MAP fault, and they asked Gordon to exchange the radio beacons, which he did, and again the command was executed without confirmation. "It's okay," Young confirmed once the engineers had made their deliberations. "You're 'Go' for docking."

Conrad had been standing off a few feet in front of the Agena, so he slipped the nose into the docking collar. "We're docked," he announced matter-of-factly. "Our docked PQI is 55 per cent."

"That's just great, Pete!" Young congratulated once again. Even although they had flown the 'brute force' rendezvous, they had made it using no more propellant than Gemini 6 had during its long and careful chase of Gemini 7. This outstanding success was another 'shot in the arm' for the Apollo planners.

Cruising along

As Gemini 11 headed out across the Atlantic on its second revolution, it undertook a new experiment. An impulse was imparted by the aft-firing thrusters and the resulting change in velocity read off the IVIs. This allowed the mass of the docked combination to be calculated, and because the mass of the Gemini spacecraft was known, the mass of the Agena could be derived. In this case, of course, the mass of the second vehicle was no mystery – the objective of the test was to determine *whether* it could be measured in this way. NASA was thinking ahead to orbital operations in which a 'space tug' would collect a payload in one orbit and drop it off in another orbit – using this technique to determine its mass to enable the requisite manoeuvres to be computed. Immediately after this test, Conrad undocked for an experiment in which the spacecraft was to manoeuvre around the Agena while a sensor measured how the charged particles of the ionosphere flowed around that vehicle.

Having served as Carnarvon's CapCom for an earlier mission, Conrad knew many of the locals. "Say 'hello' to everybody down there from me."

"Sure will," Garvin promised.

On raising Hawaii, Conrad had another request. "Ask them back there at MCC if Neil is around, and ask if he ever saw the paint all blistered off of the side of his Agena. We've got a great deal of paint off, and it looks like it's sort of anodised or something. I was wondering if during the night station-keeping we didn't spray fuel on it."

"We'll find that out for you."

When they established contact via the California relay station, Tom Stafford came on the line. "Hey, Tom," Conrad called, "did you have a lot of blisters on your Agena?"

"There were some blisters on it," Stafford recalled. In fact, of course, his Agena had been lost, and he was referring to the ATDA.

"There are some blisters on this one's TDA." Conrad was still manoeuvring alongside the Agena to finish off the ion-wake experiment.

"We'd like you to get some pictures of that," Stafford prompted. Photo-documentation would enable the engineers to figure things out at their leisure.

As they passed over Texas, Conrad announced, "We're about to record a 'first' here, the Pilot is about to dock."

"Mark one docking for Richard Gordon," the Pilot added a minute later.

"One for the right-seaters," Houston congratulated. All previous dockings had been made by the Command Pilot, sitting in the left seat.[4]

"We're cooking up a little lunch," Conrad announced when Houston checked in through Ascension.

"You're 'Go' for your PPS burn," called the *Coastal Sentry Quebec*, which was stationed in the Pacific, off Japan. A small burn to calibrate the engine's performance was to be made over Hawaii, where its telemetry could be monitored. In order not to disturb the circular orbit, this burn was to be made with the Agena TDA-south. It nudged the orbital plane by 0.2 degrees.

"John," Gordon called to Young upon making continental landfall, "riding the PPS is the biggest thrill we've had all day!"

"That baby really moves, doesn't it!"

"It was a shock, boy. It was like going off a catapult." As former naval aviators, Conrad, Gordon and Young all had experience of being catapulted off aircraft carriers.

Because the Agena's PPS had performed satisfactorily, Gemini 11 was cleared to attempt the record high apogee. Delighted, Conrad and Gordon moved to the next item on their flight plan, which was to repeat the docking in darkness to determine whether it was more difficult than in daylight.

"You guys really had a great day," Young congratulated as they grabbed supper prior to settling down for their first sleep period.

[4] All rendezvous and docking missions were to have performed a series of undockings and redockings, but Gemini 8's mission had been curtailed, Gemini 9 had not been able to dock and Gemini 10 had been low on fuel.

Gordon's EVA

"How do you feel this morning?" asked Al Bean when he awakened them 8 hours later.

"Bright-eyed and bushy-tailed," replied Conrad, but as the Public Affairs Officer pointed out to the listening journalists, the astronaut had sounded decidedly sleepy.

The highlight of the day was to be Gordon's umbilical EVA. After Cernan's experience, NASA had investigated the potential for neutral buoyancy training. After trying it, Cernan reported that the water tank accurately reproduced the inertial aspects of being weightless. However, Mike Collins had been too far advanced in his preparations for Gemini 10 to take time off to rehearse his procedures using this technique and Gordon, on hearing that Collins had achieved his objectives, had decided to follow suit. Although Collins had gone out while Gemini 10 was docked to its Agena, this had not been planned and Collins had ignored it, but Gordon's first EVA was to exploit the presence of the other vehicle.

Four hours had been set aside for EVA preparations. The checklist ran to nine closely typed pages. Conrad's attitude to a checklist was 'Get ahead, and stay ahead' on the basis that this made a margin for dealing with unexpected problems, so they were soon 30 minutes ahead of schedule. Most of their coordination was conducted on intercom, however, so as they flew over a succession of stations with only basic communications checks the PAO in Houston wryly observed that they were "maintaining pretty rigid communications silence". On finding that they were almost ready with 90 minutes to spare, Conrad briefly considered advancing the EVA, but decided to stick with the plan. Unfortunately, having transferred to the ELSS in order to test it, Gordon found himself jammed in his seat for two hours, and the heat exchanger, which was designed to operate in a vacuum, was ineffective in the pressurised cabin and he soon overheated and had to return to the ECS umbilical. Also, because they had missed a step on the checklist, Gordon had donned his gloves before affixing the outer visor to his helmet, and it took an intensive effort over a period of a half hour or so to fit it, and in addition to making himself both sweaty and irritated he cracked the visor. Starting across the United States on the final pass before the hatch was to be opened, Conrad gave their status. "We're going to hold up until we get to the next dark pass before we go any further – we're just a couple of steps off being ready to go."

Thirty minutes later, Conrad relayed through Kano in Nigeria: "We just gyrocompassed the Agena TDA-south, and we're standing by to pick up the EVA preparations again." Then over Tananarive he suggested a change of plan. He proposed rolling the Agena right through 80 degrees half an hour before the EVA was to start, and then having it maintain this attitude so that the docked combination would be at the proper angle with respect to the Sun for the opening of the hatch. Young noted that if they rolled right 16 degrees at sunrise, as planned, they would be properly illuminated when they opened the hatch 10 minutes later. "Ah, but that's not the problem, John," Conrad persisted. "The problem is – Dick will be in a 'hard' suit at that time, and he won't be able to command the Agena." The encoder was located by the Pilot's right thigh. The plan called for Gordon to roll the Agena *after* he had inflated his suit, when it would so restrict his arm that operating the encoder would be difficult. "What we'd like to do is roll right 80 degrees early. That should be the same thing, shouldn't it?"

"That's affirmative," Young conceded.

"Then that's what we're going to do," Conrad decided.

"We'd like to give you a 'Go' for depress," Bill Garvin relayed as they flew into range of Carnarvon.

"We've about five steps to go," Conrad reported, "which we'll do after we go 'Inertial'." He meant once they had rolled the Agena and commanded it to hold that attitude. With that, they resumed their preparations in silence. The roll manoeuvre was made while passing over Canton Island, but there was no communication.

Hawaii picked them up several minutes later. After examining the biomedical telemetry, Fendell called Houston. "Flight, we're showing indications that the Pilot is hyperventilating – his respiration rate is 36 per minute."

"How does the Agena look?" asked Cliff Charlesworth, the Flight Director for the EVA.

The Agena was maintaining the docked combination stable against the forces imparted as the astronauts moved around inside the cabin. "It's trying to hold an inertial attitude."

"You're 'Go' for your Stateside EVA," Fendell relayed as they dipped towards Hawaii's horizon. "Good luck."

"We have the cabin depressed," Conrad reported on establishing contact with Houston via Guaymas in Mexico. "We're standing by to open the hatch." He set the communications system to VOX so that the ground would be able to monitor Gordon while he was outside.

"I'm opening the hatch," announced Gordon. "I stood up in the seat," he later reported, "or rather, I 'flew' out, because as soon as I opened the hatch all the debris and junk went floating out – and I went right along with it." As with his predecessors, he was awed by the panoramic view through the visor. "Oh! It's a beautiful day." With Conrad holding his ankle strap to keep his foot on the seat for stability, Gordon performed his chores. After reaching back to the adapter module to release the pop-out handrail, he retrieved a cosmic ray package and passed it in to Conrad, who first tethered it in order to prevent it from floating out of the hatch and then stuffed it in his footwell.[5] Using his fist as a mallet, Gordon mounted a 16-mm movie camera on a bracket just aft of his hatch to record his activity in conjunction with the Agena. Although the preliminaries were by no means taxing, he was surprised to find himself sweating. "I've got to rest here a minute," he sighed. "I'm pooped!"

The first major task was to exit the hatch and advance to the Agena, extract the end of a 100-foot nylon tether from a dispenser on the side of the TDA and connect this to the bar on the nose of the Gemini spacecraft, which in the docked configuration was at the apex of the V-shaped guide cut into the collar. At Collins's suggestion, the

---

[5] This was the S-9 Nuclear Emulsion Experiment for the National Research Laboratory and the Goddard Space Flight Center, which was to measure the flux of cosmic rays at orbital altitude (for comparison with data from high altitude balloons). A similar package was flown on Gemini 8, but Dave Scott was unable to retrieve it.

A sequence of stills from the movie of Dick Gordon's spacewalk to fix a tether to the Agena.

umbilical had been shortened to 30 feet to make it more manageable. When Gordon was ready, Conrad played out 6 feet of umbilical. As Gordon floated forward, he did as Cernan had suggested, and used the nozzle of one of the inert RCS thrusters on the nose as a handhold prior to reaching for the collar of the TDA, but when he pushed off again he imparted an unwanted lateral component and this sent him flying across the top of the Agena.

"Did you get it?" Conrad asked.

"No, I missed it," Gordon replied. "Pull me down, Pete." Conrad tugged the umbilical to return Gordon to the vicinity of the hatch. "Easy!" Gordon warned as the impulse sent him flying past the hatch at high speed. On bouncing off the end of the line, his body began to tumble. "I can't see where I'm going." Conrad reeled in the umbilical to pull him to the hatch. Gordon set off again for the Agena. This time, he managed to grab the collar – but only by the tips of his fingers.

"Good show!" Conrad encouraged.

Gordon's next problem was to stabilise himself, so that he could extract the tether from the dispenser on the Agena and connect it to the bar. In training, he had simply sat on the thin Re-entry and Recovery section and jammed his boots in between it and the docking cone to hold himself in place. When suspended on a wire-rig trainer, it had taken only a minute or so to reach down the side of the Agena to extract the end of the tether and connect it to the bar. He had been able to do it in the weightless training aircraft, too, but in space it was much more difficult, his boots slipped out of the collar his legs floated off the spacecraft's nose.

"How're you doing?" Conrad asked, his view of the docking bar obscured by Gordon's body.

"Alright, I guess," Gordon replied unenthusiastically.

"Ride 'em, cowboy!" Conrad encouraged as he watched Gordon struggling like a bucking bronco. "Say," he prompted a minute later, on hearing huffing and puffing activate the VOX circuit, "why don't you sit down and take a rest."

"I'm tired, Pete," Gordon admitted.

"Alright, take a rest. You've got plenty of time, you've only been out 9 minutes. Take it slow." Conrad then called Houston for the first time since the hatch was opened. "Hey, you ought to see him, Houston!"

"Roger," responded Young tersely. Having observed Collins, he had a fair idea of what Conrad was seeing.

"How're you doing?" Conrad prompted again.

"I'm very tired."

"Working hard, huh?"

"Yeah."

"You look awfully funny, sitting out there in front of the spacecraft!"

On resuming his efforts to attach the tether, Gordon's heart raced at 160 beats per minute and his respiration rose to 40 per minute – twice Conrad's rate. "I think I'll take a rest."

"Take it easy," Conrad implored. Then he addressed Houston. "Dick's breathing *awfully* hard and he's resting up on front there. He's got the tether out. He's about to connect it to the docking bar." In contrast to training, Gordon had great difficulty screwing the clamp to connect the tether to the bar. Giving up on trying to keep his boots jammed in the collar, he stationed himself over the spacecraft's nose and held himself in position as best he could with one hand to counter his body's reaction to the torque as he turned the clamp using his free hand and managed to secure the clamp. On nearing the end of the Eastern Test Range, Gordon was back at the hatch resting. After Stafford's endorsement of it on Gemini 9, Gordon was to have affixed a mirror to the docking bar to enable Conrad to monitor his activities while over the adapter later on, but Conrad elected to cancel this.

"How's everything going?" Houston relayed via Ascension Island five minutes later.

"We're just resting," Conrad replied.

"Listen," Conrad called via Tananarive, "I just brought Dick back *in*. We're repressurising the cabin right now. He got so hot and sweaty that he couldn't see."

"I know how it is," Young sympathised. "When it gets where you can't see, you have got to close the lid."

Nevertheless, Conrad's recall order surprised the flight controllers. Neither Ascension nor Tananarive were configured to relay biomedical telemetry, and the Flight Surgeon had to wait for Carnarvon to assess Gordon's status. As a result of his physical exertions, Gordon's right eye had been flooded by sweat and had stung so badly as to impair his vision. Although much remained to be done, Conrad had exercised his command authority and called him back rather than permit him to move to the rear of the adapter.

Cernan had found the velcro pads and stirrups to be ineffective as stability aids, but Collins reported favourably on his handrail and McDonnell had installed more handrails and a better foot restraint to enable Gordon to gain access to the adapter and to work there. He had inherited two tasks left over from previous missions. The adapter contained a power tool that would have been tested by Dave Scott if Gemini 8 had not been cut short by a malfunctioning thruster, and the HHMU, which he was to plug into the nitrogen valve for the comprehensive evaluation that Collins had not had time to do. In effect, Gordon's exertions had overwhelmed the suit's environmental system. With the suit unable to cool him, sweat could not evaporate, and in weightlessness it simply oozed from a pore and remained in place until surface tension drew small droplets together to form larger droplets, and his eyes began to water, exacerbating the problem. The impromptu rest period had not really helped. It was no consolation that his faceplate had not fogged over. Fortunately, he was able to retrieve the movie camera, and the ingress went smoothly.[6]

"Gene Cernan warned me," Gordon reflected afterwards, referring to the effort required to maintain stability in weightlessness. "I took it to heart – I knew that it was going to be hard, but I had no idea of the magnitude."

"We're untangling all the junk," Conrad reported when Carnarvon called. By 'junk', he meant the ELSS and umbilical, which they were to jettison during the pass over the States. In doing so, they precluded the possibility of revising the second EVA to attempt to gain access to the adapter.

"Now that we've dumped the garbage," Conrad called Garvin on the next pass while they prepared a post-EVA meal, "this place looks like a grand ballroom up here."

"You're going to have a lot of fun tomorrow." If all went well, the PPS burn would raise the apogee to 750 nautical miles and facilitate an unprecedented view of the Middle East, the Indian Ocean and Australia. Carnarvon was north of Perth, right on the westernmost headland of Australia. "I don't know if you and I can survive a 23-minute pass," Garvin teased. When they flew high over Australia, they would be above his horizon for something like three times the usual duration

"We'll see who wins." Conrad had made a bet with Garvin as to which of them would run out of conversation first.

Flying high

The Canary CapCom put in the wake-up call to start the third day.

"We're up!" Gordon replied immediately.

"How're you feeling?"

"Just fine," Gordon assured. "We've been up for about 20 minutes. We're configuring for high altitude." They were packing away all the loose items in the cabin in preparation for the PPS burn.

The apogee-raising burn was to occur some 250 nautical miles west, and a little south, of Las Palma in the Canaries. As this was close to their existing perigee, this would remain unchanged by the manoeuvre, at 155 miles. The high apogee would be on the other side, some 150 miles west of Brisbane on Australia's eastern coast. They were to make two high altitude passes and re-establish their low circular orbit. The prospects for photography were excellent because the Horn of Africa was fairly clear, there was mixed cloud over the Indian Ocean, and although Jakarta was 'socked in' almost the entire Australian continent was clear.

---

[6] Apart from some footage of Cernan on Gemini 9, Gordon's EVA was the first opportunity for the planners to see how a freely-floating astronaut floundered around while trying to work.

Momentous events tend to attract VIPs, and as Canary gave the final 'Go' for the burn Robert Gilruth escorted James Elms, chairman of the Gemini Mission Review Board, into the visitors' enclosure in the control room. "The silence is so intense that you can almost hear it!" the Public Affairs Officer observed, referring to the absence of talk on the air/ground circuit.

Robert Gilruth, James Elms and Charles Mathews celebrate the Gemini 11's use of the Agena to zoom to a record apogee.

"It's going," Conrad called, reporting that the Agena's big engine was burning. "It's *really* going!"

"Cut-off," Canary informed Houston when the telemetry showed that the Agena had shut down.

"Whoop-dee-do," Conrad yelled in exhilaration, delighted that he and Gordon were now indisputably 'the fastest men alive'.

"It looked real good from here," Canary agreed. The Agena's Velocity Meter had been set to 912 feet per second. Although the tail-off had induced a 6 foot per second overshoot, this was of no great concern.

Carnarvon acquired Gemini 11 earlier than usual, as it climbed through 560 nautical miles. "Hello, up there!" Garvin welcomed.

"How long have you had us?" Conrad enquired. A ground station was usually able to read telemetry before establishing the voice link.

"About a minute."

"I'll tell you," Conrad began delightedly, "the world *is* round!"

"Have you got a good view?"

"It's spectacular, Bill," Conrad enthused. "It's fantastic! You wouldn't believe it. I've got India out the left window, Borneo under our nose and you're out the right window."

"Get some pictures out the *right* window," urged Garvin.

"We're taking them *all* out the right window," Conrad replied. The outer surface of *his* window had

As Gemini 11 climbed to its high apogee Dick Gordon snapped this spectacular view of the Indian subcontinent from a height of 400 nautical miles.

picked up a filmy contaminant during staging. Gordon was to have wiped it off with a cloth while outside, but his EVA had been cut short. The 'half moon' windows were only 6 inches tall by 8 inches wide, but a vast area of the globe was presented to them. The sensation of altitude was much more pronounced than in the standard orbit. Barely a decade earlier, the 'hot' test pilots had been dreaming of one day zooming to 100,000 feet altitude. Now, he and Gordon were thirty times higher, and still climbing. The docked combination had been oriented so that they could see the ground and, like good tourists, they were shooting pictures continuously. "I tell you," Conrad continued in awe as he looked down on the entire Australian continent, "we can't believe it!"

"For your information," Conrad noted a few minutes later as they neared apogee, "our dosimeter reads point-3 rads per hour up here." This was the 'skin dose'. The 'depth dose' was 0.11 rads. These readings were actually slightly lower than Gemini 10 had encountered flying at half of this altitude. The apogee had been positioned over Australia so as to avoid the South Atlantic Anomaly.

"Why don't you tell us about the view?" Garvin urged.

"We'll have to go on VOX while we're doing it, because we're very busy," Conrad said – he did not have time to operate the Push-to-Talk button, and so would not necessarily hear if Garvin called. "We're looking straight down over Australia now. We have the terminator out the right window." They were approaching orbital sunset. "We've the whole southern part of the world out one window! Utterly fantastic! Here comes the terminator." The signal strength diminished as they flew on, but Garvin was able to maintain contact as they headed out over the Pacific in darkness.

They were so high that the Canton Island relay station was able to pick them up almost immediately. "It sounds like you're really up there, Pete," Young mused. "Can you see New Zealand, down south?"

"No," Conrad replied, "we're past the terminator." New Zealand was 2,000 miles east of Australia and in darkness. "We're still in daylight up here, though," he added. The world below was in darkness, but they were so high that they were still catching the Sun.

"All you need is a bigger fuel tank, right!?" Young teased.

"Ha!" Conrad chuckled. If there had been a powerful Centaur available, they would have been able to shoot for a circumlunar trajectory.

Once in the Earth's shadow, they took pictures of the horizon to document the 'airglow' phenomenon which occurs in layers of varying intensity up to an altitude of about 150 miles. Collins had found that it complicated taking star sightings using a sextant. Previous missions had photographed it from just above its ceiling, but from their high vantage point Conrad and Gordon were able to observe it on a grander scale. After the perigee pass, they headed up again, and this time they turned their cameras on the Horn of Africa and the East African Rift Valley, which was one of their main terrain targets.

"How's the weather?" Garvin asked as he picked them up a few minutes later crossing the Indian Ocean.

"South of Shark's Mouth Bay they've got some cloud, but that's about it," said

Conrad, the high-flying weatherman. Because the Earth had rotated since their last pass, their apogee had migrated westward and much of the eastern part of the continent was now in darkness.

"It sounds like it's safer up there than a chest X-ray!" replied Young when Canton Island acquired them on the way down and Gordon confirmed the low dosimeter readings. Some of the managers had been reluctant to authorise flying so high, for fear of the charged-particles in the lower van Allen belt, but once Gemini 10 found the flux to be barely 10 per cent of the predicted dose, Gemini 11 had been cleared for up to 1,000 nautical miles. The radiation from the early spate of high-altitude nuclear tests that some people had expected to pose a serious risk had evidently dispersed.

"Nothing like actual data, huh!" Gordon reflected.

"Do you have any comments on the colours?"

"I'll tell you one thing," Conrad responded, "it really is *blue*. The water really stands out, and everything is blue. Obviously, the curvature of the Earth shows up a lot. Looking straight down, you can still see as clearly as orbiting down low. There is no loss of colour. Detail is good – extremely good."

Several of the 300 pictures that they took during these two high passes became 'classics' that were not surpassed until Apollo ventured out to the Moon and returned with pictures of the Earth rising over the lunar limb. As expected, the Weather Bureau concluded that its next generation of satellites would have colour cameras.

As it descended from the second high apogee, Gemini 11 prepared for the PPS burn over Texas that would recircularise its orbit. "Good burn," Gordon reported. "I took engine plume movies on all three burns." After Young and Collins had reported that the engine underwent a spectacular tail-off, the engineers had asked that this be filmed. "They should be interesting," he promised, "because the lighting conditions were different on each one."

The '1000-kilometre zoom climb' had been an exciting sight-seeing jaunt, but now there was work to be done.

Stand-up

Gordon's second EVA was to be a 'stand up' in the hatch with a routine similar to that assigned to Collins in that he was to take ultraviolet pictures of stars, but with twice as many targets spread over two periods of orbital darkness. During the intervening daylight pass, he was to take pictures of the cloud-free terrain in the continental Unites States.

"Have you thought about the window cleaning?" asked Young, as the astronauts worked through the EVA checklist.

"I don't think he can reach either of the windows," Conrad opined. As Gordon would be on the short umbilical, his 'reach' would be short. To clean Conrad's window, he would have to stretch right across the spar between the hatches and then angle down over the pane, which might put unacceptable tension on the hose. As for his own window, that would be on the far side of the open hatch. It would have been accessible on the long umbilical, but this had been jettisoned. Although there was no longer any *need* to clean Conrad's window because the high apogee photography was over, and Gordon was to snap the next sequence whilst outside, the engineers had been eager for him to at least smear a cloth over the glass to enable them to identify the residue.

"We're a little late, but I think we'll make it on time," Conrad apologised as they started across the United States for the final time prior to opening the hatch. The checklist was to be completed while over Kano, and the hatch opened while over Tananarive 12 minutes before sunset, but because they were running a few minutes late Houston did not get confirmation that the EVA had started until Carnarvon acquired the spacecraft.

"We have the camera fixed to the bracket," Conrad announced. "We're just waiting to pick up the stars – we haven't got them yet." And with that, he shut up. This time, because Gordon was on the intercom circuit, the succession of ground stations were unable to listen in. As had Collins, Gordon found working in the hatch to be straightforward. His heart rate remained about 90 beats per minute and his respiration was a steady 18 per minute, which was comparable to Conrad sitting in his seat. "It's coming along just fine. We're doing great," Conrad reported as they passed over Canton.

"That's wonderful," Young congratulated.

Whereas Collins had used exposures of 20 seconds, Gordon's ranged between 30 and 60 seconds, so he was kept busy. "We're on our last one," Conrad reported when they flew into Hawaii's range. This was fortunate, as the glow on the horizon ahead meant that sunrise was imminent.

"You might tell him," Flight prompted Fendell in Hawaii, "that I have heard there is no cloud over the Houston area this morning." He couldn't say for sure, because the MOCR was windowless.

Conrad handed out the Hasselblad for the daylight photography, and Gordon snapped the clear terrain and interesting cloud formations as they flew across the southern states. "Man! Does Houston ever look beautiful down there! Tell Dr Gilruth, I'll take his picture."

"Roger," Young acknowledged. "He'll appreciate that."

"This is not a job," Gordon reflected, "it's a privilege."

"Can you see your kids on the roof?" Young teased.

"They'd better not be!"

"Tell Dr Gilruth that we're taking a shot of the Cape, too," Gordon pointed out as they headed offshore. Having completed the terrestrial photography, he handed the camera back in to Conrad.

After informing Ascension Island that they were all set to resume the astrophotography, both men lapsed into silence and dozed off. When Young called via Tananarive, Conrad, who was on the radio circuit, was abruptly awakened. "You've got two guys taking catnaps up here!"

"Say again?" Young asked, thinking that he had misheard.

"I said we were taking a catnap," Conrad chuckled.

"That's a first," Young mused.

"We both just fell asleep a few minutes ago." As Conrad said in the post-flight debriefing, after the synoptic photography "we had the whole rest of the Atlantic to go with nothing to do so, low and behold, I fell sound asleep in my hard suit with my arms extended, and all of a sudden I woke up and realised that not only was I asleep on the job but I was asleep while we were depressurised. I said 'Hey, Dick, would you believe I fell asleep!' and all I got out of him was 'Huh, what...?', so he was asleep hanging out of the hatch on his tether and I was sitting asleep inside the spacecraft."

"Hey, John," Conrad called, remembering a discovery made by their predecessors, "how come everything floats *out* of the spacecraft? We just let little pieces of velcro go and they 'take off', straight up, out of the spacecraft, even though we're rolled over on our side."

"I think that's the 'Collins Effect'," Young retorted. "Or maybe the 'Cernan Effect'."[7] It seemed that something in the cabin was out-gassing in vacuum, and this was creating a 'wind' that transported lightweight objects out through the hatch.

"Well, it seems to work!"

When Bill Garvin announced acquisition at Carnarvon, Conrad was terse: "We're taking pictures." Then he switched back to the intercom in order to assist Gordon in completing the astrophotography.

"Hello, Hawaii!" Conrad called jubilantly a few minutes later. "We've closed the hatch and have started to repress."

"You're 'Go' on the ground," Fendell assured.

"It's a beautiful night you have down there," Conrad observed.

"I haven't had a chance to look at it, yet," Fendell admitted.

In contrast to his exhausting first excursion, Gordon later described his two and a half hours of working in the hatch as "most enjoyable".

Slowly spinning

With the hatch closed, they prepared for the tethered experiments. Conrad oriented the docked combination with its primary axis vertical, and with the Agena on the bottom of the stack. Just before flying back into range of Hawaii, he carefully undocked and eased back to draw the tether slowly out of its dispenser, but things did not go quite to plan. "We sort of upset the Agena a little bit," he told Fendell. The fact that the dispenser was on the *side* of the Agena meant that the tension in the taut tether caused the Agena to yaw, then swing back, and it was oscillating side to side with its attitude control thrusters fighting to keep it vertical.

When the tether dispenser jammed, Conrad goosed his thrusters to free the obstruction, and because the tether was attached to the docking bar above the nose of his own spacecraft this began to oscillate in pitch, which he had to try to combat as he continued to withdraw to keep the tether taut. "This is really weird," he complained. The disturbance became worse as the tether approached its full length and soaked up more energy. "Oh, man. I really upset the Agena!" He decided to cancel the gravity-gradient experiment – as this required the vehicles to be stationed one directly above the other and with the tether taut – which he did not think he could manage, and instead moved on to the rotational experiment. As they dipped towards

[7] In fact, Jim McDivitt had noted this effect while Gemini 4's hatch was open.

his horizon, Fendell reminded Gordon to use his encoder to command the Agena to turn its Attitude Control System off, for otherwise it would refuse to rotate in time with the tether. Unfortunately, the MAP confirmation was still not functioning, so they decided to wait to get confirmation from the ground that the Agena had indeed accepted the command. But the California relay station could not forward the Agena's telemetry to Houston and they were advised to hold off until they reached the Texas station – when the Agena's status could be confirmed. By the time that they were given the go-ahead to 'spin up' the tethered system, Conrad was having second thoughts. "I can't really psych out what's going on." The tether was whipping around. One moment it was taut, and the next there was a pronounced bow in it. "This tether is doing something I never thought it would do," he conceded. "It's like the Agena and I have got a skip-rope between us, and it is rotating and making a big loop."

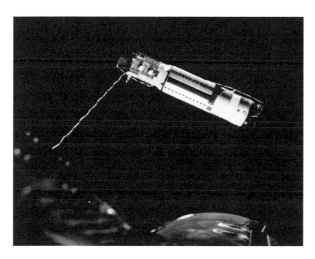

Although the idea was to maintain the two vehicles in-line with the tether taut, the dynamics took Conrad and Gordon by surprise. At this point, the tether had tugged the Agena around crosswise and the tether (attached near the TDA) is strung along the Agena's length and is in danger of wrapping around the engine.

In an effort to stabilise the Agena, Gordon commanded its Attitude Control System back on and Conrad attempted to draw the tether taut, but the dynamics persisted. "This will take somebody a little bit to figure out!" Conrad warned. The experiment was being documented by a 16-mm movie camera mounted in Gordon's window. Every time he withdrew in order to draw the tether taut, the two vehicles 'bounced' off its ends, at which time the asymmetrical forces imparted by the offset attachment points set them 'nodding', and the line went slack again as they were drawn back towards each other. "I can't get it straight." Nevertheless, by the time that they set off across the Atlantic, he had started the system rotating with a period of about 10 minutes per cycle (the objective had been for a rotation twice as fast) and while they could not *feel* the resulting centrifugal force its effect could be noted by placing an object against the window and watching it drift. They flew on through the night in this state. When Hawaii reestablished contact, Fendell passed on Houston's recommendation that rather than attempting to cancel the rotation they should jettison the docking bar and hastily move clear of the Agena, which they did, leaving that vehicle inverted and with the tether horizontal.

The experiment had been frustrating, and had consumed rather more fuel than expected, but the cut-off had been a PQI of 10 per cent, and because they had twice this remaining the Flight Dynamics Officer proposed a re-rendezvous of a type that had never been tried before. Because the spacecraft's radar was inoperative (it was still in the anomalous state that it had assumed during the later stage of the initial rendezvous) radar tracking by the WWTN would be used to navigate the spacecraft to a point from which its crew would be able to brake using optical tracking. In preparation, Conrad returned to the Agena and nulled the relative motion to achieve, as Young put it, "a good solid station-keeping position", and then he performed a prograde/up separation manoeuvre to enter an elliptical orbit with an early apogee. As they climbed above and therefore fell behind the target, they watched it race across the backdrop of South America.

"Boy!" Gordon exclaimed. "It really moves along the ground. No wonder Tom and Gene had trouble."

"Now you *know* you're going 17,500 miles an hour," Conrad reflected. "Rendezvousing from over the top must have been fantastic!"

When the *Rose Knot Victor* in the South Atlantic picked them up, they were already "a couple of miles" above the Agena. At the intersecting node about 75 minutes after separation they were to perform a retrograde burn in order to resume the Agena's orbit, to station-keep a few miles in trail of it. To Conrad's surprise, they were requested to yaw around 180 degrees and use the *aft* thrusters to make this burn. When he asked why, he was told that it was a follow-up to the experiment in which the spacecraft had driven the Agena forward to assess the scope for measuring the mass of a 'cargo module'. As the Program matured, Houston was starting to 'load up' the flight plan. In this case, it was exploiting a programmed manoeuvre to calibrate the thrusters. Conrad and Gordon rounded out their most exciting day with a cold supper and went to sleep. Because the cancellation burn had not been exact, during the night they slipped about 25 nautical miles behind the Agena.

FIDO's triumph

The re-rendezvous objective was to use WWTN radar tracking to manoeuvre to simulate arriving at the apogee of a Hohmann transfer orbit, at which time the crew were to brake by optical tracking. This was to be achieved by descending several miles, to catch up with the Agena in such a way that as the range reduced the relative velocity approximated that at one of the midcourse corrections of the canonical 130-degree transfer. One option was to descend sufficiently far to yield a fast catch-up and make the interception early on, but this would not provide much time for tracking. The other option was to establish this relative velocity much later, in order to allow the radars time to determine the ideal time to initiate the braking phase. Since it was to be initiated from the same orbit as the target, the Flight Dynamics Officer called it a 'stable orbit rendezvous', but in fighter pilot slang it was to be a 'ground-controlled interception'. Young carefully explained this improvisation to the crew. "I want to make clear that the primary purpose of the intercept manoeuvre that you are about to do, is to evaluate the *ground vectoring capability* – we want you to use your computer only for the intercept initiation manoeuvre," Young began. As for their part in the action, they were to dig out the charts carried to enable them to proceed with a nominal 130-degree transfer in the event of a radar-failure. Houston would supply a 'phantom' TPI point and what would have been one of the transfer corrections – as if they had actually made that TPI burn – and then they were to scale down the results to allow for the fact that they had *not* made it. "You understand that the only midcourse correction you'll have a chance for is as you break out into sunlight there, towards the end." They assured him that they understood. In effect, they were simulating the final phase of a fuel-efficient 'tangential' rendezvous. As this technique was very sensitive to set-up errors, lighting conditions in the braking phase were difficult to control. Their chances of success were dependent on having the propellant to recover from any errors that became evident as they made the interception. Young warned that they were to cancel the braking and fly past the Agena if the PQI fell to 2.5 per cent.

As they flew out across the Atlantic, Conrad pondered an ambiguity in the instruction to "divide by 3" the velocity given by the charts for the midcourse correction. Unless they knew precisely what they were to do, they risked a collision. "Regarding the nominal range-rate at that midcourse correction," he began when they reached Canary Island, "do you want us to divide *that* by 3, too?" And then, by way of clarification, he added: "I'm just trying to see what we're going to be closing at."

"Just tell him to go into the charts with the nominal numbers as they would for a regular rendezvous, and then just divide the delta-V answer by 3," Charlesworth told the CapCom, who relayed the information.

"I *understand* that," Conrad persisted. "But what I want to know is will my closing rate be one-third of what is on these charts?"

"Standby one," the CapCom advised.

"I don't think we know the answer to that," Charlesworth admitted.

When Houston established a relay through Kano, Young reiterated the procedure. "You don't divide the angle by 3, it's just the delta-V that you calculate."

"What I'm interested in, John," Conrad tried once again, "is what will our *closing rate* be?"

"I think it would be pretty close to nominal-divided-by 3," Young ventured.

"In other words, it's nothing we can't hack looking out the window, without the radar?" Conrad posed.

"It'll be pretty slow," Young assured, "like 15 feet per second."

Finally satisfied, Conrad made a retrograde/down burn to enter an elliptical orbit with its perigee 5 nautical miles below the Agena's orbit, so that after subtending a central angle of 258 degrees, as Gemini 11 began to climb again, its trajectory would *both* mimic the ascent to the apogee of a Hohmann transfer and the point in a canonical transfer 34 degrees prior to the braking point. If they managed to brake, they would draw alongside the Agena some 292 degrees from their starting point. Houston's willingness to try such a rendezvous indicated its mastery of orbital operations. The sequence was initiated approaching Australia. "How'd the burn go?" Garvin asked upon establishing contact.

"Just fine," Conrad said matter-of-factly.

With a PQI of 9 per cent, they had a fair chance of success so long as their trajectory was reasonable. As they passed over Australia, a radar at Woomera checked the result of the burn. When Houston picked them up half a revolution later, they were nearing perigee and about 4 nautical miles below the Agena's orbit. "We're going to get some tracking over the States, and try to give you an estimate of what the 'correction' should be," Young advised.

Gordon was to calculate this final burn independently on the basis of his optical tracking. Conrad sought clarification of whether they were to give priority to the ground's version.

"It'll be in the ballpark," Young opined ambiguously.

In the Earth's shadow, Conrad tried to track the Agena by its lights but he had a problem. "My window is so greasy," he warned Young, "that I can't see him through the reticle. I'm using the reticle with my left eye and tracking him with my right eye." The resulting parallax effect undermined the principle of the reticle. "The grease spot is right in front of the reticle."

"You'll be able to see him at sunrise," Young assured, "grease or no grease!" The Agena would be blindingly bright.

"He's coming out into the sunlight right now!" This time though, he was prepared for the glare and slipped on his sunglasses.

The 'correction' was to be made 6 minutes later, over the North Atlantic. As they reached the end of the Eastern Test Range, Young offered Houston's best estimate: 6 feet per second forward and 2.4 feet per second to the right; Gordon's solution came out the same, which was encouraging.

"Have you made your midcourse?" the Canary CapCom asked when he picked them up.

"Affirmative," replied Conrad.

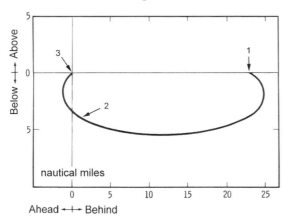

How Gemini 11 performed the 'stable orbit' re-rendezvous with its Agena target vehicle.

1. Start interception
2. '34-degree correction' after 1h 5m
3. Visual TPF after 1h 13m

Gordon began by using the rate of increase of pitch to calculate their range and range-rate, but in light of Collins's experience on Gemini 10 he switched to estimating its angular size in his sextant once they were within 8,000 feet. At 1,000 feet they were closing at 15 feet per second, and Conrad began to brake. While they had been away, the gravity gradient had flipped the Agena vertical, engine down, and the tether was now trailing straight upwards.

"We're home free!" Conrad called down delightedly. "We're just sliding in there – very peachy."

"Wonderful!" Flight Director Glynn Lunney repeatedly issued over the network circuit. As a former FIDO, he understood the celebration in The Trench. The interception orbit had been so accurate that the spacecraft had consumed only 45 pounds of fuel. Chris Kraft had joined Lunney, to witness the outcome. NASA was certainly showing that it knew how to fly in space.

"Everybody in Houston's real happy with that, 'Eleven," Canary pointed out.

"They're not any happier than I am," Conrad assured. "Or Dick."

No sooner had they drawn alongside the target than Houston ordered them home. Conrad described the Agena as "the best friend we ever had". It had performed all of the manoeuvres assigned to it with perfection and, with the exception of the MAP light, had given no trouble. "It was very kind to us," Gordon agreed.

"If you guys'll send a tanker up," Conrad had joked a little earlier, "we'll stay up a little longer and do some more work." Naval aviators, he and Gordon were used to extending their sorties by in-flight refuelling. Young had retorted, "The tanker's on the water."

As they prepared to return to Earth, the Gemini 11 crew were able to content themselves with the knowledge that they had significantly advanced the state of the art of rendezvous.

Closed loop

By this time, NASA was willing to try a 'closed loop' re-entry, so instead of following the computer's steering cues by 'flying the needles', Conrad was to monitor its performance and be ready to intervene if it strayed off course.[8] It successfully corrected a cross-range error and then delivered the spacecraft within 3 miles of the *USS Guam*. Although not as good as Gemini 9's remarkable performance, this was the most accurate descent to date.

"This old world looks pretty good from the deck of a carrier," Conrad reflected

after they had been recovered, "but boy, it sure looks great from 750 miles out." His verdict? "We had a good flight."

Al Shepard (right) welcomes home Pete Conrad and Dick Gordon.

---

[8] In fact, the computer aimed for the centre of an elliptical 'footprint' on the ocean that was 300 miles long and 30 miles across.

Chapter 10

'THE END'
Gemini 12

Seeking a mission

Early in 1966, when it had been hoped to launch the Apollo 'shakedown' mission in the autumn, tentative plans were made for the final Gemini mission to rendezvous with Apollo 1 to symbolically 'pass the baton' in the race for the Moon, but the Apollo/Gemini Mission Planning Coordination Committee rejected this as a distraction.[1]

On 17 June 1966, after they had backed-up Gemini 9, Deke Slayton formally named Jim Lovell and Buzz Aldrin to Gemini 12. At this point, as Lovell subsequently reflected, they did not really have a 'mission', they knew only that they would be called upon "to catch all those items that weren't caught on previous flights". Actually, if the Apollo launch had not been postponed to 1967, Gemini 12 would very likely have been cancelled. One school of thought was that it was being flown simply because the hardware had been paid for. For a while, it seemed that the rendezvous part of the flight plan would have to be deleted due to the lack of a target. Of the six GATVs built, five had been launched, of which two had been lost, and the ATDA that had been added as a contingency had already been used. The only option for Gemini 12 was to refurbish the first production unit, which had been assigned to ground testing.[2] However, as the 'spare' Atlas had been used with the ATDA there was no launch vehicle. When an Atlas was 'borrowed' there were calls for the mission to be cancelled because most of the Program's goals had been accomplished. In July, the expectation was that Gemini 12 would make an 'm=2' rendezvous, but the Gemini Mission Review Board revised this in late September to 'm=3'.[3] Whilst docked manoeuvres were to be made, there was no plan to chase a passive Agena. George Mueller dismissed as a distraction an early suggestion to rendezvous with the Orbiting Astronomical Observatory that had been crippled when its power system blew a fuse shortly after being launched in April. It was finally decided to use the Agena to enter a high orbit for photography, as on Gemini 11, but this time with the 400 nautical mile pass over the United States.

---

[1] The Apollo/Gemini Mission Planning Coordination Committee had been established in January 1965.

[2] In its refurbished state for Gemini 12, the initial Agena production item was redesignated as GATV-5001R.

[3] The 'm=1' rendezvous had just been demonstrated by Gemini 11; 'm=3' by Gemini 9 and (allowing for the navigation experiment on the first orbit) on Gemini 10; and 'm=4' by Gemini 6 and Gemini 8. The 'm=2' case was under consideration for a lunar module rendezvousing with its mothership.

Atlas vehicles and GATVs

SLV-5301	GATV-5002	Assigned to Gemini 6 but lost when its PPS exploded
SLV-5302	GATV-5003	Used by Gemini 8
SLV-5303	GATV-5004	Assigned to Gemini 9 but lost when the Atlas malfunctioned
SLV-5304	ATDA	Used by Gemini 9
SLV-5305	GATV-5005	Used by Gemini 10
SLV-5306	GATV-5006	Used by Gemini 11
SLV-5307	GATV-5001	Used for ground tests and then refurbished for Gemini 12

Jim Lovell and Buzz Aldrin innocently emerge from the suit-up trailer on Pad 16 and walk up the ramp on Pad 19, each with a card on his back saying 'THE' and 'END', respectively, signifying the final mission of the Program.

Back to basics

When Aldrin was assigned, about the only certain item on the flight plan was an AMU test because the agreement with the Air Force called for two flight opportunities, and as he had served as Cernan's backup, and therefore was familiar with the system, it was logical that the test should be assigned to him. Following Cernan's favourable report on the fidelity of neutral buoyancy simulation, Aldrin employed it as his primary training aid. It was certainly an improvement over the wire and pulley rig, and by providing prolonged periods of pseudo-weightlessness it facilitated 'integrated simulations', which were not practicable flying in the KC-135 aircraft. As a result, he was confident of being able to egress and move slowly and deliberately to the adapter to don the backpack. But on 23 September the Gemini Mission Review Board concluded that Collins's successful retrieval of the package from the inert Agena had been misleading, and that the difficulties suffered by Cernan and Gordon suggested that the likelihood of Aldrin being able to test the AMU was low, and to Aldrin's dismay the AMU was deleted from the mission. Given the state of the art, this was the sensible decision – NASA had no wish for the Program to draw to a conclusion with another 'failed' EVA.

Nevertheless, the Board acknowledged that EVA was the one Program goal that had yet to be mastered, and recommended that the *primary objective* of this final flight should be to perfect EVA techniques. George Mueller accepted the report, and asked Aldrin to develop a two-hour "investigation of EVA fundamentals through repetitive performance of basic easily-monitored and calibrated tasks." The most important lesson to be drawn from the previous outings was that whilst standing in the hatch bestowed sufficient stability to do useful work, activities beyond the hatch that required maintaining a given position were difficult because every physical act involved applying a force, which induced a reactive force. If there was a conveniently located handrail, it was possible to maintain position with one hand and work with the other but not all tasks could be accomplished single-handedly. If the HHMU proved capable of providing the *mobility* to reach a work site, it was apparent that the outstanding issue was the *stability* to work, and if doing so required two hands then stability would have to be achieved by some other means, such as by 'standing' in foot restraints. The long-term impact of this realisation was that *every* spacewalking task would have to be thought out in advance, and appropriate stability aids installed. McDonnell greatly increased the number of hand and foot restraints[4] and added facilities for Aldrin to use waist-tethers to fix himself in place at a worksite so as to leave both hands free to work. Also, for the first time, aids were developed for the Agena. Aldrin devoted the rest of his underwater training to testing moving purposefully around the two vehicles, and determined the optimum placement for the various aids. The individual tasks for this 'back to basics' EVA seemed at first sight to be trivial, as indeed they were on Earth, but the objective was to provide *leverage* in weightlessness. By the time he set off, Aldrin was so much better prepared than his predecessors that if he was unsuccessful in space, NASA would have to rethink its future plans for using astronauts to outfit spent stages as space stations and to assemble large orbital structures.[5] The flight plan was not settled until 20 October. It followed the Gemini 10 model, in which the first EVA would be done in the hatch.

---

4 Cernan's spacecraft had been provided with a total of 9 restraints; Aldrin's spacecraft had 44 restraints.
5 On 6 August 1965 George Mueller had formed the Apollo Applications Program to plan all manner of tasks for spacewalkers.

This reflected the belief that this initial exposure would serve to acclimatise Aldrin to working in weightlessness, and better prepare him for more demanding activity on going outside the following day.[6] In fact, an additional day was assigned to enable him to undertake a record-breaking three EVA's – the final one being another 'stand up'.

By combining a demanding EVA programme with a high apogee to document the entire continental United States in a single picture, Gemini 12 promised to be a memorable finale to the Program.

Gemini 12 lifts off.

A tricky rendezvous

The planned dual launch on 9 November was cancelled in order to investigate a problem with the Atlas's autopilot, but on Friday, 11 November, everything was ready. Lovell and Aldrin arrived at Pad 19 wearing small cards taped on their backs, one saying 'THE' and the other saying 'END'. On entering the White Room they found a large banner proclaiming 'Last Chance. No Relaunch. Show Will Close After This Performance'. A stack of paper one-foot thick, labelled 'Flight Plan', was presented to Lovell as a joke. In fact, if the Atlas were to fail to insert their Agena into orbit, Gemini 12 would launch anyway and proceed with the EVA assignments, but it lifted off on time at 14:08 and performed flawlessly. Although there was a momentary dip in the Agena's thrust chamber pressure, it went on to attain the desired orbit. Gemini 12 left at 15:46, further confirming the competence of Colonel John Albert's Gemini Launch Vehicle Division of the Air Force's 6555th Aerospace Test Wing. No sooner had the Titan left the pad than a salvage team moved in to strip it of anything useful, as a preliminary to decommissioning it.

"You're steering right down the pipe," advised Pete Conrad, the CapCom in Houston as the second stage steered into the target's orbital plane.

"Man!" Aldrin exclaimed. "This is a pretty good visual simulation."

"How about the horizon, Buzz?" the veteran Lovell asked his rookie right-seater. "Beautiful."

The Titan left them 24 feet per second short, so Lovell rectified this with the IVAR burn, which incorporated the first part of the out-of-plane correction. "Right on the schnozz," he observed to Aldrin on the intercom. As they left the Eastern Test Range behind, they worked on the post-insertion checklist.

---

[6] A follow-on to this line of thinking was that if Gordon's two excursions had been switched around, he may have been more successful during the full egress.

Conrad tried to call via Ascension Island, but failed to elicit a response. "Do you have a comm problem?" he asked on raising them via the Tananarive relay station on the island of Madagascar. "Or was it us?"

"We didn't hear a thing," Lovell reported. In fact, due to a scheduling mix-up, Ascension had not been 'on air'. Aldrin had tried to make contact but was not surprised by the silence because they had barely entered the station's communication circle. Before he too lost them, Conrad read up the data for the manoeuvre that was to raise their perigee from 87 to 119 nautical miles in order to establish the phasing for the rendezvous.

"Look at that sunrise, Buzz," Lovell observed as they approached the dawn terminator. "It'll come up and zap our eyeballs any minute now." He put on his sunglasses.

"We're in the process of burning," Aldrin announced when Jim Fucci in Carnarvon made contact.

"Do you have some information on the out-of-plane burn?" Lovell enquired.

"You're 'Go' for the onboard solution," Fucci replied.

On reaching Hawaii, Lovell told Keith Kundel they had "burned the residuals right down" on the plane change.

"We have a solid lock-on, now," Aldrin informed Houston when they approached the US mainland.

"It looks like the radar meets its specs," Conrad mused. The radar had locked onto the Agena's transponder at the record range of 236 nautical miles. Although he left them in peace as they started across the continent, he called when they reached Texas to read up the rest of the rendezvous solution derived from radar tracking. A tweak was made on the second revolution, just beyond the Eastern Test Range. Ascension was another low-elevation pass, so he had to wait for Tananarive to find out how they were doing.

"We got visual contact through the sextant at a range of 85 miles," Lovell managed to report before losing contact.

The co-elliptic manoeuvre to put the spacecraft in an orbit 10 nautical miles below that of the target was made several minutes before reaching Carnarvon. As Aldrin then prepared the computer to monitor the radar during the ensuing catch-up phase, he saw that the range-rate was nonsensical. In fact, the lock-on light had gone out as well. The signal strength could be seen to fluctuate on the analogue meter, and the computer could not handle the intermittent signal. "We seem to have lost our radar," he told Carnarvon.

"Standby," Fucci advised. He checked the spacecraft's telemetry. "We show you locked-on." He checked the target's transponder. "The Agena is receiving you." He then left them in peace, but as they dipped towards his horizon he asked how they were doing.

"Still no lock-on," Aldrin reported. Deciding not to rely on the radar recovering in time for the computer to calculate its 'closed loop' solution for the TPI burn, he retrieved his backup charts and began to track the target optically using his sextant.

"We're proceeding with a radar-failure rendezvous," Lovell announced upon contacting Hawaii.

To gain some insight into the nature of the problem, Kundel requested that Aldrin use his encoder to change the brightness of the lights on the Agena's Status Display Panel while he monitored the telemetry to determine whether the command was accepted and executed – it was, but the MAP light in the spacecraft did not light. Lovell recommended that they try a different antenna. Kundel concurred. Aldrin issued the command and the telemetry showed that the Agena switched antennas, but it made no difference. With the nominal time for TPI looming, Aldrin, the rendezvous specialist in the Astronaut Office who had helped to develop the backup procedure, was sure that he would be able to calculate the manoeuvre himself.

"We've got a visual on him now, both with the sextant and in the reticle," Lovell reported, "and it looks – from our angle – as though we're going to be a little bit *later* than nominal." At the requisite pitch angle, Lovell made the burn to start the transfer.

"I don't know what you're reading in the cockpit," Conrad said as they started across the United States, "but it sure doesn't look too good on the ground."

Aldrin noted that although the radar was still intermittent, it occasionally got a valid range – the range-rate figure, however, was always nonsense.

"The acquisition lights stand out very nicely against the stars," Lovell remarked. The fact that it was fixed indicated that their trajectory was excellent. Aldrin's first transfer correction was very small, and the second was so insignificant that it was clear that they would make it providing they managed to brake properly.

"Have fun!" Conrad called, as they headed back out over the Atlantic.

On the nominal rendezvous, they would have waited for sunrise before starting to brake, but as they slipped within 10,000 feet Lovell slowed to 17 feet per second. "Closing well," Aldrin assured. At 4,000 feet Lovell slowed to 10 feet per second, and at 1,000 feet to 5 feet per second. A year earlier, losing the radar/computer interface prior to initiating the terminal phase would have been a 'show stopper', but NASA had now mastered orbital manoeuvring.

"We're here!" Lovell announced when Conrad established contact through Tananarive. In fact, they were still 500 feet out and slowly braking, with the Agena in their spotlight. A few minutes after they drew to a halt, the Sun rose. Lovell manoeuvred to enable Aldrin to check the Status Display Panel.

"You're 'Go' to dock," Conrad called, but they did not hear him. To fill the eight-minute gap to the next contact, Lovell manoeuvred to enable Aldrin to document the condition of the Agena for the engineers.

"Hey, how're you doing up there?" asked Bill Garvin, the CapCom on the *Coastal Sentry Quebec* stationed south of Japan.

"Everything's fine. We've got about 67 per cent PQI," Lovell replied.

"Is he docked?" Glynn Lunney demanded.

"That's negative, Flight."

"Are you satisfied with the Agena?"

"Both vehicles are 'Go', Flight."

"Let him know that."

"Gemini 12, *CSQ*. We're giving you a 'Go' for docking."

"Roger," Lovell acknowledged. "Thank you."

"Let me know when he docks," Lunney ordered.

Lovell, however, continued his fly-around until Aldrin ran out of film and then he lined up and eased the nose into the TDA. "We're docked."

On the way to Hawaii, Aldrin commanded the Agena to yaw the docked combination to perform the stress test of the docking collar.

"Have you finished gyrocompassing?" Kundel asked when they rose above his horizon.

"We're Agena 0-0-0," Aldrin replied. They had completed the manoeuvre with the Agena TDA-forward. The next item on the checklist was a series of undocking and redocking tests. Accordingly, Lovell withdrew and began another fly-around.

"Are you letting Buzz fly that thing now?" Conrad asked as they started across the US mainland.

"He'll have plenty of opportunities," Lovell assured.

"He hasn't used up his share of the fuel yet," Aldrin pointed out.

"Yes, watch him," Conrad advised, "he's a 'time' hog."

"We're right around the tail of the Agena," Lovell announced a minute later, "looking up the engine."

At this point, Conrad relayed ominous news. "Jim, we saw an anomaly in the PPS when it was being burned into orbit, and we're still going over the data right now, but the problem appears to be one that indicates a possible turbine pump problem." He was referring to the fluctuation in the thrust chamber pressure.[7] The engineers were debating whether it would be safe to reignite it. "We're going to give you a 'Go' a little bit later as to whether you can make the PPS burn or not."

Lovell accepted this stoically. If the engine was unsafe, there was nothing they could do about it, and in any case their primary objectives were EVA-related. "Has my wife had that baby, yet?" he enquired as they flew down the Eastern Test Range on their fourth revolution.

"Yeah, I think so," Conrad congratulated. Marilyn Lovell had given birth to Jeffrey, their fourth child and second son.

Having passed into darkness, Lovell switched on his docking light and prepared to redock. When the *Rose Knot Victor* in the South Atlantic briefly picked them up, the telemetry was intermittent and Lovell said that they were suffering a slight control problem, but flew out of range before he could be more specific.

"Were they undocked for your *whole* pass?" Lunney asked of the ship's CapCom.

"That's affirmative."

Lovell elaborated a few minutes later via Tananarive. "We had a little difficulty at night. We attempted a docking and we thought we lost our roll control, so we backed out." In fact, as he drove his spacecraft's nose into the Agena's collar the two vehicles had not been aligned and only one of the three latches had triggered, inducing an unexpected motion and jamming the spacecraft in the TDA. To break free, he had had to repeatedly fire the fore/aft thrusters, giving rise to an awful grating vibration. Their ungraceful withdrawal had upset the Agena, but its Attitude Control System had restabilised it. However, with their PQI now down to 56 per cent Lovell recommended

---

[7] It was concluded that a faulty bearing was causing the fuel turbopump to overspeed.

deleting the remaining docking tests.

"We concur," Conrad agreed immediately.

A few minutes later, Lovell changed his mind. "We're contemplating another docking to try out our control system again."

Conrad had been thinking back to his own docking trials. "When you undocked, did you use your manoeuvre thrusters to undock with?" The spacecraft's manoeuvre thrusters were for translating and were much more powerful than those used for attitude control.

"Affirmative."

"Okay, listen, just send the command, and the TDA will shove you out." The TDA was spring-loaded.

"That's an idea," Lovell agreed.

"We'd like you to do it over the *CSQ*, which is your next station," Conrad relayed. "Just back out to assure yourself that your control system is okay, and then redock. We'll give you a 'Go' for that over the *CSQ*. We'd like to look at you a little bit when you do it."

When Garvin picked them up, he passed the word: "Both vehicles look good. We'd like you to go ahead and undock – and we'd like you to do it *slowly*."

"Let's get this undocking underway," Lunney instructed impatiently when there was no news.

"Go ahead and undock," Garvin urged.

"We're going to undock as soon as we can," Lovell promised. His attention was focused on diagnosing a warning light.

"How does it look?" asked Flight.

"He's still docked. Do you want to hold off until Hawaii?"

In fact, because Houston wanted the telemetry monitored during the undocking/redocking there was no option but to wait. "Let's hold off."

"Hold off until you get to Hawaii," Garvin relayed.

"Roger," Lovell acknowledged just before he flew out of range.

When Gemini 12 reached Hawaii, Lovell was ready. "We'll commence the undocking."

"We've got an indication of spacecraft-free," Kundel reported Houston.

"Undocking was successful," Lovell reported.

"How's it look to you?" Lunney demanded.

"It looks good, Flight" Kundel replied. "Are you satisfied with the control system?" he asked Lovell.

"The control system is 'Go'." They had just been unfortunate in setting up their previous manoeuvre; it was embarrassing but no harm had been done.

With the issue resolved, Lunney told Hawaii to relay the bad news. "Tell them about the PPS."

"You're 'No-Go' for the PPS burn," Kundel relayed. "Instead of doing that, we're going to have you do an SPS retrograde burn to set up the solar eclipse."

The Gemini 12 patch was inspired by the intention to make the first orbital observation of an eclipse of the Sun.

Chasing the Moon's shadow

If Gemini 12's launch had not slipped, the mission would have been able to accommodate *both* observing a solar eclipse and a high-apogee photographic pass across the United States, but the delay had prompted a choice and the eclipse had been sacrificed. Whilst the high orbit was impractical without the Agena's big engine, the less powerful Secondary Propulsion System could be used to set up the phasing required to pass through the Moon's shadow. When they yawed around for this burn, Lovell was surprised that the Agena overshot by 40 degrees and its Attitude Control System took a while to line it up properly, and even then the vehicle had a tendency to roll. The likelihood that the rapid turn had left the propellants sloshing around in its tanks raised concern in Houston about the vehicle's stability during the burn.

"If she gets away from you," Conrad warned earnestly, "take over with the OAMS."

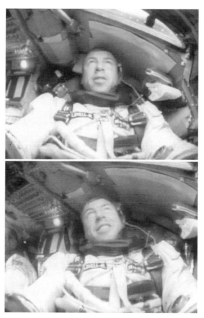

Jim Lovell inside.

If this had been the first mission to attempt a docked manoeuvre using the Agena, NASA would probably have cancelled the burn, but the Flight Director gave the go-ahead at Lovell's discretion.

"How did the burn go?" Garvin asked once the *Coastal Sentry Quebec* picked them up.

"We burned the Agena." It had maintained its orientation to within a few degrees.

"Does the vehicle look alright?" Lunney asked.

Garvin checked the telemetry from the Agena. "It looks good."

Lovell and Aldrin had supper and retired for the night. On taking over from Lunney, Gene Kranz revised the flight plan for the second day in space to take into account the fact that the PPS was not available, and Bill Anders read it up immediately after awakening the crew. The eclipse was top of the list. After almost 20 minutes of traversing the penumbra section of the Moon's shadow, during which there would be little to do, the plan called for feverish activity during the 8 seconds of totality, because they were to shoot three 'stills' at various exposures in the hope of documenting the solar corona around the lunar disk and then yaw the docked combination around to document the Moon's shadow racing over the face of the Earth using a 16-mm movie camera. On the original flight plan, Aldrin was to have opened the hatch to photograph the eclipse using ultraviolet film, but this observation had been deleted and could not be done through the window.

"How did the eclipse photography go?" Anders enquired afterwards.

"We hit the eclipse right on the money," Lovell reported, "but we were unsuccessful in picking up the shadow." They had yawed slowly to preclude upsetting the Agena, and by the time that they had turned to face the Earth the spot of the Moon's shadow was over the local horizon. The next item on the flight plan was Aldrin's first EVA.

First EVA

For his first excursion Aldrin was to open the hatch just prior to sunset and stand in it to undertake ultraviolet astrophotography during two night passes and various other tasks in the intervening period of daylight which – if they had been able to use the Agena's PPS – would have included synoptic photography from the high apogee over the continental United States. One experiment required Aldrin to exercise for 30 seconds by vigorously swinging his arms to move his hands between his waist and his helmet. He was to do this twice: the first time after depressurising the cabin and whilst still seated, and then again in the hatch at some convenient time. The telemetry from his heart and respiration monitors would enable the metabolic cost to be computed and compared with preflight tests.

"'Twelve, you're 'Go' to depress on schedule," said Anders as they flew overhead.

"Do you have anything you want to add to our jettison bag?" Lovell asked.

"That's negative."

"Everything looked good here, going over the hill," Flight Director Cliff Charlesworth assured Canary Island as they left the Eastern Test Range. By the time Canary picked them up, Lovell had put the communications system on VOX to enable the ground to eavesdrop on the intercom. Aldrin was already in the hatch. When he threw the canvas bag of trash over the side, it narrowly missed the Agena's tall dipole antenna.

"I believe I see stars all around us," Aldrin announced. He was surprised, because it was still daylight.

"Are you sure it isn't the depress?" Lovell thought it more likely that the 'stars' were ice particles from the cabin depressurisation reflecting the Sun.

"I don't think so." But then Aldrin realised that he had been fooled. "Wait a minute! That 'star' was the trash bag!" Moving on, he compared his actions and reactions in space with his experience in the water tank and decided that he was in a *familiar* environment – making rapid movements had been difficult against the water resistance so he had learned to make slow and deliberate moves, and doing so now in space was second nature. "Do you notice any motions of the spacecraft combination when I move around just a little bit?" Aldrin enquired.

"Yes, a little bit."

When they entered the Earth's shadow, the 'stars' which were ice crystals alongside the vehicle winked out and the real ones blazed into view. Aldrin started to take pictures for the astrophotography experiment with exposures ranging between 30 seconds and 2 minutes, using a cable release to operate the camera's shutter. When Carnarvon called he reported that he had slipped "way behind the timeline – I don't know how they expect us to do it."

"Gemini 12, Houston," Anders relayed when they flew within range of Canton Island.

"Stand by," Lovell advised, "I can't do anything now."

Anders tried again a minute later.

"We're on a tight schedule right now," Lovell insisted. Several minutes later, however, he came back. "We're in daylight again. Tell the experimenters that we couldn't catch Gamma Velorum because we didn't have time."

Aldrin dismounted the camera and handed it to Lovell. The first set of photographs were taken through a diffraction grating, and the second set were to be taken through a prism. The diffraction grating would provide a greater spectral resolution, but the prism would enable fainter objects to be recorded for a given exposure. While Lovell set about reconfiguring the camera, Aldrin turned around in the hatch to face the rear and retrieved the micrometeoroid package from the adapter immediately aft of his hatch. He then retrieved a telescoping pole from a stowage fixture on the inside of his hatch, extended it, poked one end into a hole in the Agena's collar and attached the other end to a fixture just aft of the spar between the hatches. The pole was to act as a handrail to provide access to the Agena during his second excursion for the 'back to

Standing in the hatch Buzz Aldrin installed a telescoping pole to enable him to make his way rapidly and efficiently to the Agena.

basics' programme. When Lovell handed him the daylight camera, he started synoptic photography. Oceanographers had asked for pictures of ocean currents, eddies, river outwash and plankton blooms, so he snapped these while over the Pacific. On starting across America, he switched to terrain photography.

"How's the weather in Houston," Lovell asked Anders.

"It's cloudy – we have a front laying just off the coast."

"Yeah," confirmed Aldrin , "I think I can see it." There were only scattered showers from southern California to Texas, but the southeast was generally socked in.

"What did I tell you, Buzz," Lovell teased. "Four days of vacation with pay to see the world!" To Lovell this short event-packed flight was a delight compared to a fortnight cooped up in Gemini 7 gathering biomedical data. "It's a pity we can't be a little higher." If they had been able to make the high-apogee pass, Aldrin would have had a coast-to-coast view of the continent. Leaving the Eastern Test Range behind, they headed out over

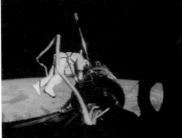

A series of stills from a movie of Buzz Aldrin working on the Agena. In the first frame, just before he egressed, he has his visor raised, and in the final frame he is displaying his homage to Veterans' Day.

the Atlantic for yet another circuit of the Earth.

"Here's Africa coming up," Aldrin noted. It would soon be dark again. As he manoeuvred to hand his camera in through the hatch, his boots strayed onto one of the instrument panels.

"Watch it!" Lovell warned. "Get your foot out of the way – you're on the switches to the fuel cells."

"I don't want to hit those!" Aldrin confirmed. Once deactivated, a fuel cell would not be able to be restarted, and the loss of one of the fuel cells would probably curtail the mission. Lovell could not see all of the switches to verify that they had not been disturbed, so Aldrin had to shuffle to peer down to check them and in doing so upset the straps that were holding him in position. "I don't know if I'll be able to get these straps on, the way the suit is."

"If you can," Lovell suggested, "get in the same position you were in before, and I'll put the strap in the same spot it was before."

By the time they reached Tananarive they were in darkness again and Aldrin was working methodically through the list of stars. To his surprise, repeatedly holding the cable release for the two-minute exposures made his fingers ache. "When I rub my gloves together," he noted in fascination, "there is static electricity between them." Once his eyes had fully adapted to the darkness, he had noticed that his gloves glowed. Experimenting, he found that rubbing his thumb against his index finger induced an electrostatic effect, evidently resulting from passing through the ionosphere – in effect, he was flying through a sea of electrons.

When Garvin reminded Aldrin to do his second period of exercise, he said that he was too busy. As before, the schedule proved overly optimistic and he ran out of time; he took the last few pictures as the Sun was rising. By Canton Island, he had resumed his seat and closed the hatch.

This 'stand up' had confirmed the conclusion from Geminis 10 and 11 that working in the hatch was undemanding. Working methodically through tasks that were manageable, Aldrin's heart had hovered at around 75 beats per minute and his respiration rarely exceeded 18 per minute – he had been almost as relaxed as Lovell, who was hardly stressed at all. After lunch, Lovell and Aldrin spent the rest of the day on experiments, including photographing 'airglow' and the sky in an effort to discover any small rocky bodies that might be trapped in the two gravitationally neutral libration regions, 60 degrees ahead of and behind the Moon in its orbit around the Earth.

'Back to basics' EVA

The highlight of the next day was to be Aldrin's full-egress EVA. The hatch was to be opened at sunrise, ten minutes after leaving Carnarvon. After unstowing the 25-foot umbilical and plugging it into the ELSS, he waited until 30 minutes prior to hatch-opening before going on the external system in order to avoid the overheating suffered by Gordon. A 'clothes line' tether had been run across the cabin, and items that would be used on the EVA had been tied to it to preclude them floating out of the hatch.

"Have fun," Anders called as they passed over Kano. Their ground track did not provide another opportunity to communicate until Corpus Christi in Texas, some 25 minutes later, by which time Aldrin had set up a movie camera behind his hatch facing forward to record his activities on the Agena, and had translated down the 6-foot handrail that he had installed the previous day. He had two short tethers on the waist of his suit's webbing harness, and these had hooks at the ends. On reaching the TDA, he hooked the left tether to the loop at the end of the handrail and the other to a ring on the docking bar, positioning himself as if 'swimming' with his head above the TDA and his feet near the open hatch – a considerably more relaxed posture in which to work than the 'cowboy' stance astride the vehicle used by Dick Gordon on Gemini 11. Aldrin effortlessly retrieved the end of the tether from its dispenser on the side of the TDA and affixed it to the docking bar. He then exposed a micrometeoroid package that was to be left for possible future retrieval. In a scheduled rest, he attended to a personal item. "To commemorate our launch day on November 11, I have an emblem here that I'd like to leave in orbit – it reads: 'November 11, Vets' Day'." In fact, if their launch had been on time, this EVA would have taken place on Veterans' Day.

"Stay right there, Buzz," Lovell said, as he prepared to take Aldrin's picture. "Hold it a second."

"I'd like to extend the meaning of it," Aldrin continued his ceremony, "to include all the people in the world who have been, and are now, and will continue to strive for peace and freedom."

"Mighty fine," Anders agreed.

Aldrin then sprang a surprise on his commander. "I've got another one here." He held up a second emblem. "This message concerns a contest coming up in the future." He meant the classic football game between the Army's West Point and the Navy's Annapolis. "I think the precedent was set a year ago." He was alluding to Gemini 6's rendezvous with Gemini 7, when Wally Schirra, an Annapolis graduate, had displayed a 'Beat Army' sign in his window to tease Frank Borman, a West Point man. Lovell was Navy, too. "I'm not sure that Jim can read this one," Aldrin, another West Pointer, continued, "but I'll read it out loud to you so you can all hear it – 'Go Army, Beat Navy'."

"Roger, '*Beat Army*'," chuckled Anders who graduated from Annapolis prior to joining the Air Force.

"I knew we had the wrong CapCom on today," Aldrin mused. He set the emblems adrift.

"Okay, Buzzeroni, how do you feel?" Lovell enquired when it was time to resume work.

"Great!" Aldrin assured. He then gave a running commentary as he made his way along the rails to the rear of the spacecraft's adapter without difficulty. As predicted, standing in the hatch on his first EVA he had been able to 'calibrate' the forces required to start moving purposefully, slowly, and without imparting a rotation. "I'm putting one hand over the other, Jim, just getting myself going with a little momentum." After he "turned the corner", he was in the cavity at the rear of the adapter. He clamped his umbilical in a tool so that it would not flex and disturb him. "Both lights are working." Unlike Cernan he would be able to see what he was doing while the spacecraft was in the Earth's shadow. Once he had flipped himself the 'right' way up and was facing 'forward', he grasped two rails and eased his boots into the 'overshoe' restraints which were sturdier than the stirrups that had been provided for Cernan. These were painted gold to reflect sunlight in order not to damage his boots, and so had been dubbed his 'golden slippers'.[8] "My boots don't seem to have any tendency to come out," he continued his commentary. "But the heel seems to be a little higher than expected." With both boots anchored, he was able to release his grip on the rails and let his body adopt the suit's 'neutral' posture, hunching him over slightly, with his arms out in front of his chest. "Now, if I move a little forward toward the workstation, I have a very small tendency to bounce back, but I know that I can put the suit in several different positions and I'll stay in place." It was clear that he would be able to maintain a fairly comfortable position in order to perform his tasks on the 'workstation'. "I'm going to lean back now," he continued. "I'm leaning straight back." In this position, with his boots in the restraints, he had his knees bent and his spine aligned with the vehicle's axis. "This is a little harder than it was underwater; it requires a bit more leg force." With the vehicle flying inverted, he was looking straight down; there was tan desert below – Africa.

"We're going to start the adapter workstation tasks," Lovell reported when Kano picked them up. "How do you like the foot restraints, Buzz?"

"They're great." By this point, Aldrin had tied himself to the workstation using his waist tethers.

"Three minutes to sunset," Anders warned. The 'workstation' tasks were scheduled to start at sunset. Aldrin had positioned himself with time to spare and, in marked contrast to Cernan at this point, his heart was relaxed.

"We're going to the checklist now," Lovell warned. He was to call out each task in order, and log the time that it took Aldrin to do it. The aim was to evaluate the degree of difficulty associated with each task, undertaken in a variety of conditions. Because the tasks were not interdependent, if one task proved to be unexpectedly

---

[8] Gemini 11 had similar 'overshoe' restraints, but Dick Gordon had not had the opportunity to assess them.

time-consuming he was to abandon it and move on – his purpose was to gather data. And in addition to formal rest periods, he had the option of calling a time-out if he felt tired. In all, he had 17 numbered assignments which were deemed to be representative of tasks likely to be required on space station missions. In particular, he was to test a torque wrench and measure the force needed to tighten and then release bolts.[9] He also had a special torqueless wrench.[10] In order to compare one-handed with two-handed tasks, he was to put plugs into sockets, some of which simulated electrical connectors and others fluid connectors. He was also to sever short lengths of electrical cable using a pair of shears. To assess dexterity using suit gloves, he was to link together sets of small hooks and loops. As an additional leverage test, he was to assess the force required to tear off pre-positioned strips of velcro of different widths.

"What's the status of your visor? Is it fogged?" Lovell checked.

"It's as clear as a bell."

"Are you perspiring at all?"

"Negative."

When Carnarvon acquired the spacecraft 10 minutes later, Aldrin was commentating on his progress with one of the bolts. "... A loose bolt with a washer is being inserted manually; this is a delicate operation. ... The bolt is just fitting straight into the hole. ... 0-g is holding it there; it's not engaged. ... That would have been a beautiful picture!" The 16-mm movie camera he had installed to record his activities had refused to wind film. "I fumbled the bolt and the washer. They went drifting in underneath my helmet. I pushed them forward, then moved myself away for a moment, caught both of them, and put them together and they're now going in manually."

"You're playing a little orbital mechanics to retain proficiency?" Lovell teased.

"I had to do a little rendezvous." The fact that Aldrin had been able to ease his body out of the way and retrieve the two items, one in each hand, more or less simultaneously, served as a convincing demonstration of the effectiveness of the new foot restraints. "I'm using the wrench to tighten it up," he continued, having inserted the bolt manually. "There's a problem overcoming the ratchet – in other words, you unwind most of what you've just wound!" He was producing useful data for the engineers.

"Let me know when you have it tightened down," Lovell reminded. He was trying to time this activity.

Next, Aldrin performed his dexterity test by effortlessly linking together the hooks and rings, and then disassembling them again.

"Now take a 'banana pill' and rest for a while!" Lovell ordered. Because his workstation tasks had been derided as 'monkey work', the backup crew had put a yellow *Chiquita* sticker beside his workstation. Although Aldrin had not remarked upon it, Lovell was in on the joke.

---

[9] These bolts were similar to those used in the 'dome cap' of the S-IVB stage, which astronauts would have to release if they were to enter a spent stage and convert it to a 'wet' laboratory.

[10] The torqueless wrench had been assigned to several missions, but until now nobody had managed to test it.

By the time they reached Canton Island, Aldrin was making his way back to the Agena, to perform some similar tasks on a workstation there. This time he tethered himself with his head facing Lovell's window and had his legs projecting out over the Agena. On completing his tasks, he threw the loose items over the side.

"Boy, you're a litterbug, aren't you!"

"Alright, the workstation is clear."

A close up of Buzz Aldrin alongside the Agena.

"Hey, Buzz," Lovell said, as Aldrin started back, "you never did wipe off my window, did you?"

"Oh, okay. Give me half a minute."

"Would you change the oil too," Lovell quipped.

"Is that helping any?" Aldrin asked as he wiped the window.

"No."

"It's slippery. Wait a minute, I think it's coming off." Finally, NASA had a sample of the contaminant. "See any improvement?"

"Yeah, it looks good. I know where you can get a job, Buzz." Lovell was recommending Aldrin as a gas station attendant.

"I can see from the outside," Aldrin observed as he examined the window close up, "that you have quite a film on the *inside*." The fact that the contamination was between the inner and outer panes suggested that it was due to out-gassing in the space environment rather than a coating picked up during the ascent through the atmosphere. If this was the case, then the windows of the Apollo spacecraft would probably have to be redesigned – as it would not do to arrive in lunar orbit and be unable to see out!

Once Aldrin had retrieved the movie camera, Lovell pulled most of the umbilical into the cabin and Aldrin resumed his comfortable stance in the hatch. Because they had been having trouble with one of the OAMS thrusters, Lovell asked him to observe while he fired it. "I can see something coming out," Aldrin noted. "How about hitting another one for comparison." Lovell fired a healthy one. "That was a lot cleaner flame."

"Was the first one a flame, or a fluid spewing out?" Lovell asked.

"There was barely a flame on either of them that I could see." Lovell fired the ailing one again. "There's quite a difference," Aldrin confirmed. "It looks like there's a lot of unburned material coming out." This was a useful observation, as one of the technical objectives of the Program was to find out whether the thrusters could work for at least the time that an Apollo lunar mission would require, but on several missions some of them had degraded.[11]

As his final task, Aldrin detached the 6-foot-long telescoping handrail and threw it away.

"You're the world's highest javelin thrower!" Lovell laughed.

As Aldrin resumed his seat, his 'back to basics' EVA had been as satisfyingly successful as some of his predecessors' experiences had been frustrating. Ironically, if the Air Force's AMU had been in the adapter, he would have been able to don it without difficulty, but with this being the final mission of the Program there would now be no opportunity for the 'Buck Rogers' act.

In the gravity gradient

After repressurising, they had a snack and prepared for the gravity-gradient test left over from Gemini 11. Conrad was on the CapCom console in Houston to offer advice. On making contact with Carnarvon, Lovell said that he had the docked combination stable in a vertical orientation, with the Agena beneath. After Carnarvon, he waited for sunrise, then undocked and eased slowly back to start to draw out the tether. In a brief link via Canton Island, Aldrin reported that the Agena was "standing there nice and vertical". By the time that they reached Hawaii, the 50-foot tether was fully deployed, although not yet taut.

"We'd be in great shape right now if we had a control system!" Lovell informed Kundel. "We're hardly moving, except that we can't control attitude." Whenever he tried to yaw or pitch up to cancel the excursions imparted by the tether on the docking bar, the fact that two of the thrusters were not working meant he induced an anomalous roll component. On making contact with Guaymas in Mexico, he explained that he was concerned that the tether might foul the Agena's antenna. As they crossed the continent, the orientation of the tethered 'system' slowly migrated, tracing out an arc of a circle. Having begun directly over the Agena, they had first drifted ahead of it and were now off to one side. Whenever the tether went taut it

---

[11] The thrusters were protected by an ablative material, and if they got too hot this burned off.

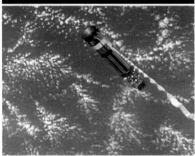

A series of stills from a movie of an attempt at forming a stabilised system using the Earth's gravity gradient by linking two vehicles with a tether.

gently drew the vehicles back and relaxed again. The Agena's Attitude Control System was holding it stable, but Lovell had decided to play possum and let his spacecraft loll around. As they set off across the Atlantic, he decided that because the tether was fairly taut, it appeared that the system was settling down and it would be safe to continue in darkness. At sunset, he switched on the docking light to monitor the state of the tether. On establishing contact via Tananarive, he said that he had decided to try to reposition his spacecraft over the Agena. He was still manoeuvring when Carnarvon picked them up, but his efforts were rewarded and he was able to tell Hawaii that the system was oriented vertically, the tether was taut, and they had ceased bobbing off its end. When Kundel asked if he believed they had been caught in the gravity gradient, Lovell was enthusiastic: "It sure looks like it!" With the line taut, the system oscillated slowly with excursions 60 degrees to either side of vertical, but it was evident that they had indeed been captured by the gravity gradient. As they flew on through another night they grabbed another bite to eat.

"We can get off the tether any time you want," Lovell said on re-establishing contact with Hawaii, but the response was to make another night pass on the tether and release the Agena at sunrise because Houston wanted a separation burn about 30 minutes later, and given their thruster problem this was to be made in daylight to ensure that they would be able to react to any attitude excursions. The tether was released by jettisoning the docking bar to which it had been connected. "We're released," Lovell reported via Tananarive. The separation manoeuvre was just before reaching Hawaii. As it was to be a prograde burn, Lovell oriented Gemini 12 backwards and used the forward-firing thrusters so that they could keep an eye on the Agena as they departed. They still had a PQI of 34 per cent, but the thruster problem had prompted Houston to dismiss the option of a re-rendezvous.

Having said goodbye to the Agena, Lovell and Aldrin had their supper and retired for a well-earned rest.

Buzz Aldrin snacks during the tether experiment.

Final EVA

"Ready to depress," Lovell relayed to Anders via Kano. Aldrin opened the hatch while crossing the Indian Ocean, tossed out the accumulated trash – mostly food packaging – and stood on his seat. On flying into darkness, he resumed the ultraviolet astrophotography work where he had left off on his first EVA. Given the thruster issue, it was impractical to mount the camera on the bracket and have Lovell turn the spacecraft to aim the camera at each star field for long exposures, so Aldrin held the camera and shot 1-second exposures. As the Sun began to brighten the horizon ahead, he snapped a sequence of pictures for a scientist with a desire to study dust in the atmosphere, then lowered his protective visor and resumed the astrophotography even although he could no longer see his targets, until the film was finished. He had offered to stay out and take some pictures of the United States, but the offer had been declined, and so as they ran up to the California coast he ingressed and shut the hatch, having been out for barely an hour. He had been totally at ease. His heart rate had hovered at about 85 beats per minute and his respiration was a relaxed 16 per minute. It was just another day at work. Over an accumulated five and a half hours of external activity, he had convincingly 'ticked the box' for EVA, and with it the last of the Program's primary objectives.

Home

After lunch, they set about photographing weather systems, interesting terrain features, airglow on the horizon and the lunar libration zones. When Lovell discovered that two more thrusters had worn out he had to resort to 'free drift', which restricted the opportunities for photography. They continued to drift as they slept, and the next morning Aldrin assessed a new sextant being considered for Apollo, to verify it could make accurate measurements. In the final hours of the mission the output of the fuel cells declined dramatically. Nevertheless, it was decided to stick to the plan and return to the prime recovery zone, and so the batteries were brought on line early to pick up the load.

As Gemini 12 descends on its parachute a helicopter hovers to retrieve Jim Lovell and Buzz Aldrin to take them to the recovery ship, and a lone swimmer remains with the capsule until it can be hoisted onboard.

Jim Lovell and Buzz Aldrin are welcomed onboard the *USS Wasp*. Note the satellite dish for the 'live' TV transmission via the Early Bird satellite.

Like Conrad, Lovell let the computer fly the re-entry. This proved fortunate because as they descended they were twice distracted. As Aldrin held his Hasselblad up to the window to document the spectacular plasma stream, it slipped and smashed onto his chest. Moments later, a storage bag that had been velcroed to the sidewall broke free and ended up in Lovell's lap, and he jammed it between his legs to prevent it from striking switches. The computer's steering was impeccable and it clipped several hundred feet off the 'closed loop' record. The sea-state made the splashdown a little rough, but within 30 minutes the helicopter had picked them up and set them down on the flight deck of the *USS Wasp*. The 'live' television broadcast was projected onto the screen in Mission Control and prompted, as the Public Affairs Officer put it, "a tremendous ovation". The Gemini Program was finished! The mood was a mixture of delight that it had finally achieved *all* of its primary objectives, relief that the ground teams would finally be able to take a break after non-stop operations, and eagerness by those who had already moved on to get the Apollo 1 mission underway.

Chapter 11

Apollo
Meeting Kennedy's Challenge

Gemini Legacy

NASA had flown ten manned missions since Alexei Leonov made his historic 'walk' in space, but no cosmonauts had flown during this time. Although the Soviets would clearly have learned by observing Gemini, they had not gained practical experience.

The Gemini Program not only demonstrated that rendezvous and docking in space were feasible, by evaluating a variety of techniques it gave the Apollo planners the flexibility of options. Although the 'm=3' method was selected for the early lunar landing missions, this was later superseded by the much more demanding 'm=1'.

Gemini showed that astronauts could endure the space environment for longer than any Apollo mission would require. Given that the longest American space flight at the time of John F Kennedy's commitment to Apollo was Al Shepard's 15-minute suborbital arc – on which he had been weightless for only a couple of minutes – this was welcome news. The fuel cells that were to power the Apollo mothership were tested on Gemini, as was a fully inertial reference platform for guidance and navigation, a spaceborne radar, a state-of-the-art digital computer to process the radar data for rendezvous, and the hypergolic bipropellant ablative thrusters. Gemini not only tested spacecraft technology and trained astronauts, it also developed a variety of ground support equipment such as simulators and trained flight controllers to give Apollo a running start. And by establishing that astronauts could operate outside their craft, Gemini paved the way for a rescue option for the crew of a lunar module that was unable to dock with its mothership. Gemini established that a spacecraft could be steered through re-entry to splash down at a given target for rapid recovery. The achievement of controlled re-entry illustrated the Program's methodology. On the early missions, the issues were worked out. Astronauts then showed that they could 'fly the needles' and follow the computer's cues to make accurate splashdowns. Only then was the computer allowed to fly autonomously, but even then the crew monitored its performance, ready to intervene if it malfunctioned. The development of controlled re-entry increased NASA's confidence in the 'atmospheric skip' that an Apollo mothership returning from the Moon was to use to bleed off energy prior to re-entry.[1]

---

[1] Apollo 8, which was the first to attempt such a re-entry, splashed within 2.4 nautical miles of the target, and Apollo 10 trimmed this to less than a mile.

#	Ocean	Recovery Area	Recovery Ship	Error (n.m)
Gemini 3	Atlantic	4-1	*Intrepid*	60
Gemini 4	Atlantic	63-1	*Wasp*	44
Gemini 5	Atlantic	121-1	*Lake Champlain*	92
Gemini 6	Atlantic	17-1	*Wasp*	6.97
Gemini 7	Atlantic	207-1	*Wasp*	6.38
Gemini 8	Pacific	7-3	*Mason*	-
Gemini 9	Atlantic	46-1	*Wasp*	0.38
Gemini 10	Atlantic	44-1	*Guadalcanal*	3.35
Gemini 11	Atlantic	45-1	*Guam*	2.65
Gemini 12	Atlantic	60-1	*Wasp*	2.60

Source: *On the Shoulders of Titans*, SP-4203, NASA 1977

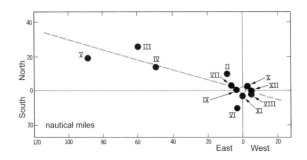

This splashdown plot clearly shows that the technique of controlled re-entry was mastered.

The wisdom of using Gemini to develop re-entry techniques for Apollo contrasts with the Soviet strategy of sending unmanned versions of its lunar spacecraft looping around the Moon in order to test the 'skip' manoeuvre. Only one of a series of half a dozen tests was completely successful.[2] Of course, the success rate may have been significantly better if a cosmonaut had been onboard to take over when the automated system suffered a problem. Ironically, these on-going difficulties were used to overrule calls for manned missions.

Run up to Apollo

In 1962, after being 'pulled' from his Mercury flight for medical reasons, Deke Slayton was appointed Assistant Director of Flight Crew Operations, and gained responsibilty for selecting the crews. In May 1963 Gordon Cooper flew a 34-hour mission that stretched the capability of the Mercury capsule to its limit. Nevertheless, Al Shepard, who had made the first suborbital arc on a Redstone in 1961, lobbied to ride an Atlas for a three-day Mercury mission in the autumn of 1963, but his efforts

---

[2] Although the tests were made by a stripped down version of the Soyuz spacecraft, these missions were called Zond in order to disguise their true purpose.

came to nothing.[3] Gus Grissom, in contrast, acknowledging that he had no chance of making a Mercury orbital flight, had switched to the Gemini Program, where he played a key role in the development of the new spacecraft. The other 'Original Seven' astronauts were not available. After 'wasting' propellant on his Mercury flight, Scott Carpenter's prospects were dashed by senior management and he had taken a leave of absence to return to the Navy to pursue deep sea research. John Glenn had announced his intention to retire in order to pursue a career in politics.[4]

In the summer of 1963 Slayton told Shepard and Tom Stafford that they would fly the first Gemini mission, with Gus Grissom and Frank Borman backing them up. By October it had become evident that the development of the Agena target vehicle was running late. There was no option but to slip the Program's rendezvous objective and devote the early missions to the endurance objective. In August 1963 Shepard had been tentatively diagnosed as having Ménière's syndrome, an inner ear ailment that induced vertigo. When this was confirmed in March 1964, Shepard was grounded. Grissom moved into the prime crew with John Young, backed up by Wally Schirra and Stafford. This was announced on 13 April, a week after the successful Gemini 1 test, and in was not long before Schirra leaked to reporters the fact that he was to command the first rendezvous mission.[5]

In July 1964 Slayton assigned Jim McDivitt and Ed White to the first endurance flight, which at that time was to run seven days, with Borman teaming with Jim Lovell in backup. However, a few months later it became evident that the fuel cell system that was to power the long-duration missions was also running late, and it was decided to assign McDivitt a spacecraft equipped with the maximum number of storage batteries that it could reasonably carry, and fly it to its limit, which was four days, and to form a new crew of Gordon Cooper and Pete Conrad to tackle the seven-day mission when the fuel cells became available. When the schedule had been drawn up in July, the second unmanned test was set for August and it was hoped to fly Gemini 3 in December. However, a variety of delays meant that Gemini 2 could not be dispatched until 19 January 1965. Gemini 3's successful 'shakedown' flight on 23 March slipped the Program into gear, and on 15 April NASA announced that Schirra and Stafford would fly Gemini 6 and attempt the first rendezvous.

In addition to giving rookie astronauts on-the-job training in backup roles, the 'rotation' system enabled an astronaut to cycle through a backup assignment after a flight in order to pass on his experience. Right-seaters stood a chance of being promoted to the left seat for a second mission, but a left-seater would either transfer straight to Apollo or fulfil a backup slot on his way to Apollo. Talking to reporters in the run up to Gemini 6, Grissom, now in backup, was asked if he expected to fly another Gemini. He replied that he did not think so, but expected Young to get a

---

[3] James Webb dismissed it as a wasteful diversion of funding from Gemini.

[4] In fact, President Kennedy had told NASA that Glenn was a national icon and they should not risk losing him in another space mission.

[5] Slayton established a 'rotation' in which (generally speaking) a backup crew would skip two missions and then fly the next one.

commander's slot. Asked for his view, Young replied, "I hope Gus is right."

After their Gemini 4 mission in June 1965, President Johnson sent McDivitt and White to the Paris Air Show, where they met Yuri Gagarin. On 1 July, Slayton assigned McDivitt to Apollo to follow the development of the lunar module, as a preliminary to commanding the crew that would test it in Earth orbit. White was rotated to be Gemini 7 backup commander with Michael Collins.[6] After Gemini 5 finally wrested the endurance record from the Soviet Union, Cooper and Conrad were sent on a tour that included the International Astronautical Federation in Athens, where they met Pavel Belyayev and Alexei Leonov. The backup crew was split up, and Slayton assigned Neil Armstrong and Elliot See to command Gemini 8 and Gemini 9 respectively, with Dave Scott and Charlie Bassett as their right-seaters. This was because after White's difficulty in closing the spacecraft's hatch following his spacewalk, it had been decided that only the more muscular astronauts would be assigned EVAs. Conrad rotated to backup Gemini 8 with Dick Gordon, which put them in line for Gemini 11. After Gemini 6, Stafford rotated to backup Gemini 9 with Gene Cernan, which set them up for Gemini 12; Young rotated to Gemini 10 with Michael Collins; and Lovell was assigned to backup Gemini 10 with Buzz Aldrin. After Gemini 8, Armstrong rotated to backup Gemini 11 with C.C. Williams. After Gemini 9, See was to rotate to backup Gemini 12 with Al Bean. As the Program reached its half-way point, therefore, the line up for the remaining missions was clear.

This plan was disrupted by the deaths of See and Bassett in an air crash on 28 February 1966. Slayton promptly advanced the backup crew to prime – this was the first time that it had been done – and thus within months of returning from Gemini 6, Stafford was preparing for a second mission.[7] Faced with the need for a new backup crew, Slayton advanced Lovell and Aldrin from this function on Gemini 10. This opened the way for a reshuffle of the rest of the crew assignments. Firstly, advancing Lovell and Aldrin gave them the opportunity to rotate to Gemini 12. Whilst Cernan rotated to backup, he retained his right-seat status so as to enable him to focus on assisting Aldrin in EVA training. Slayton appointed Cooper (who had otherwise fallen out of the rotation) to serve as the commander of this dead-end slot. To fill the backup slot on Gemini 10, Slayton paired Bean (who would, in any case, no longer be able to fly with See) with Williams. Armstrong retained his slot as backup commander of Gemini 11, but with newcomer Bill Anders. As events transpired, the deaths of See and Bassett would have a major impact on the Apollo crewing.[8]

Meanwhile, in early 1966, Grissom was told that after backing up Gemini 6 he

[6] Michael Collins was the first of the Group 3 astronauts to receive a flight assignment; although he was not to be the first to make it into space – this distinction fell to Dave Scott.
[7] Tom Stafford became the first astronaut to fly two Gemini missions. If it were not for the loss of the original Gemini 9 crew, this distinction would have gone to John Young on Gemini 10.
[8] One example being that Buzz Aldrin would almost certainly not have been eligible for Apollo 11, and would not have been a member of the first crew to descend to the lunar surface.

The crews for the Gemini missions were drawn from the first three groups of astronauts.
At the top (seated) is the group recruited in 1959 for Mercury, with the
second (1962) standing behind them; below is the third (1963) group.

would be given command the first test of the Apollo mothership. After Gemini 4
McDivitt was made Grissom's backup. After backing up Gemini 7, White was
assigned to Grissom's crew as the 'Senior Pilot'. Rookie Roger Chaffee was assigned
as 'Pilot'. On returning from the round-the-world tour after Gemini 6, Schirra was
given command of the second Apollo flight, with rookies Donn Eisele as Senior Pilot
and Walt Cunningham as Pilot. Slayton told Borman to backup Schirra, and
earmarked Bassett and Anders as his crewmates. After Gemini 8, Scott was made
Senior Pilot on McDivitt's crew, and rookie Rusty Schweickart was assigned as Pilot.
The loss of See and Bassett disrupted Borman's nascent crew, so Slayton reassigned
Anders to a Gemini backup slot. Borman picked up Stafford as Senior Pilot after his
return from Gemini 9, and Collins as Pilot after Gemini 10.

The Flight Directors also cycled through to Apollo. Chris Kraft moved to Apollo after Gemini 6/7, John Hodge joined him after Gemini 8 and Gene Kranz followed after Gemini 9, in order to provide three shifts.[9]

The Block I version of the Apollo CSM had been designed prior to the decision for the LOR mission mode, and because it had no docking facilities it could not operate with a LM. At first it was thought that there would be a whole series of Block I missions, but by early 1966 only two remained and they were to be launched on Saturn IB rockets and fly 14-day missions to shake out the bugs. Slayton's only rule for crewing these missions was that the commander was required to have prior experience. In early 1966, when he started to create crews, it was expected that the first mission would fly before the end of the year. However, the development of the Block I had fallen behind schedule. Grissom's spacecraft, CSM 012, was to have been shipped from the North American Aviation factory in Downey, near Los Angeles, in early August, but a problem with a glycol pump in the Environmental Control System (ECS) delayed its shipment to 25 August. On its arrival, John Shinkle, the Apollo Manager at the Cape, was quick to note that it would require a great deal of work to make it fit to fly. The replacement of the ECS held up vacuum chamber tests until mid-November. Although on 7 October the Design Certification Review Board said that CSM 012 "conformed to design requirements", Shinkle drew up a lengthy list of outstanding "major" problems involving the ECS and the primary and attitude-control propulsion systems. When a propellant tank in CSM 017 ruptured during a pressure test on 25 October, Joe Shea, the Apollo Program Manager, grounded CSM 012 until this fault had been identified, in case it was a generic problem. However, an investigation revealed that the methanol used for the test as a propellant substitute (the actual propellants were toxic) had reacted with the titanium and induced stress corrosion; the tanks in CSM 012 were certified fit to fly by repeating the test using a non-reactive fluid.   In early December, NASA finally acknowledged that CSM 012 would not be ready for launch until February 1967. The spacecraft was mated with its Saturn IB in early January, and its launch set for 21 February.

With the development of the Saturn V falling behind schedule, the mission planners were inspired by the dual launches in which a Gemini had lifted off as its Agena target completed its first orbit, and it was decided to mount dual missions in which a LM would be launched on a Saturn IB, with the CSM following on another such rocket one orbit later. The first of these missions was to be flown in mid-1967, after the two Block I CSM tests. For missions with a LM, Slayton added a rule that the Senior Pilot must have experience of rendezvous, because if the LM were to become disabled the CSM would have to rescue its crew.

In late October, Schirra started to lobby to have the second 14-day mission in a Block I deleted, arguing instead that he be assigned the first test of the Block II CSM with a LM. In November the Block I reflight *was* cancelled, but Slayon pulled

---

[9] They left Cliff Charlesworth and Glynn Lunney, both newly promoted off the MOCR floor, to handle the final three Gemini missions. In fact, Lunney also managed the first Saturn IB launch for Apollo in early 1966.

McDivitt out from backing up Apollo 1 and gave him the CSM+LM test, and he made Schirra backup Grissom on the 14-day mission. Why? Because Eisele did not have experience of rendezvous whereas Scott did. With the cancellation of the second 14-day flight, Borman's backup crew was split up once again. Stafford was given command of the crew that would backup McDivitt's mission, with John Young as Senior Pilot and Gene Cernan as Pilot – in the process forming a highly experienced crew that could be rotated to one of the first missions to venture to the Moon, and possibly even attempt the first landing. As 1967 began, therefore, NASA was confident that it would be able to meet Kennedy's challenge of landing a man on the Moon by the end of the decade.

Gus Grissom, Ed White and Roger Chaffee, who were to have flown Apollo 1.

Setback

On 26 January, Schirra's crew performed a 'full up' systems test of CSM 012 in which everything was powered up, but the power was drawn from the ground and the capsule was not pressurised with pure oxygen. It had not been a very productive day. "Frankly, Gus," Schirra said in the debriefing with Grissom and Shea, "I don't like it. You're going to be in there with full oxygen tomorrow, and if you have the same feeling I do, I suggest you get out." But there was a determination to push on and catch up on the several-times-delayed schedule. The next day, Friday, 27 January, Grissom, White and Chaffee set out to conduct the final systems test demonstration, a 'plugs out' test in which the spacecraft would be on internal power and pressurised with pure oxygen at 16 psi (that is, slightly above ambient) for an integrity check. After a simulated countdown, the day was to end with an emergency egress drill.

In Houston Flight Director John Hodge was monitoring progress, but the action was at the Cape. Slayton was in the Pad 34 blockhouse talking to Rocco Petrone, the Director of Launch Operations. Stu Roosa was maintaining communication with the spacecraft, as the 'Stoney' CapCom. Five miles away, in the Automated Checkout Equipment facility of the Manned Spacecraft Operations Building, 'Skip' Chauvin was serving as the Spacecraft Test Conductor.

"Fire!!" yelled Grissom at 18:31 Cape time, in a hold at T–10 minutes. "We've got a fire in the cockpit."

In all, there were 25 technicians on Level A8 of Pad 34's service structure, and five more either on the access arm or in the White Room. Henry Rogers, NASA's Inspector of Quality Control, was in the elevator, ascending the service structure. Systems technician L.D. Reece was waiting for the 'Go' to disconnect the spacecraft for the 'plugs out' test, which had been delayed by problems with communications, notably the whistle from an 'open' microphone that could not be located.

"Get them out of there!" yelled Donald Babbitt, North American Aviation's Pad Leader, on hearing Grissom's call. Mechanical technician James Gleaves was closest, but a spout of flame burst from the capsule before he could react, and he was beaten back by the flame and smoke.

Gary Propst, an RCA technician, was on the first floor of the pad structure, monitoring the feed from a TV camera in the White Room that showed the window in the spacecraft's hatch. On hearing Grissom's call he looked up and saw the brilliant light in the window with gloved hands moving about within.

As soon as Slayton realised what had happened, he sent medics Fred Kelly and Alan Harter to the pad. "You know what I'll find," Kelly observed pointedly. The best that they would be able to do would be to supervise the retrieval of the bodies. On reflection, Slayton decided to accompany them. "We were the first guys from the blockhouse to reach the pad," he later noted. Despite the intensity of the fire, Grissom, White and Chaffee had died by asphyxiation as a result of the toxic fumes created by the incomplete combustion of the synthetic materials in the spacecraft. Although they received second and third degree burns, these would not in themselves have been fatal. After only a few minutes, Slayton left the White Room to call Houston to explain the situation. Shea had just arrived back in Houston and was briefing George Low when the news came through.

The Astronaut Office in Houston was very quiet. All the 'old hands' were absent. With Slayton away, Don Gregory, his assistant, ran the routine Friday staff meeting. The meeting had only just convened when the red phone on Slayton's desk rang. Gregory answered, then reported, "There has been a fire in the spacecraft." Michael Collins was the senior astronaut present. Collins arranged for Bean to track down the wives. In each case, the news had to be broken by an astronaut who was also a close friend of the family. Charles Berry and Marge Slayton went to see Betty Grissom. Conrad was sent to find Pat White. Cernan would have been ideal to notify Martha Chaffee, because he lived next door, but he was in Downey with Stafford and Young working on CSM 014, so Collins went to give her the bad news.

Shepard was in Dallas, about to make a speech at a dinner. He was taken aside and told about the fire. Schirra, Eisele and Cunningham were in the air returning from the Cape, and were told upon touching down at Ellington Air Force Base, which was effectively Mission Control's local airfield. Schirra called Slayton at the Cape, who filled him in on the details.

James Webb, Robert Seamans, Robert Gilruth, George Mueller, Kurt Debus, Sam Phillips and Wernher von Braun were at the International Club in Washington D.C. with Gemini and Apollo corporate officials – including Lee Atwood of North

American Aviation – to mark the transition from Gemini to Apollo. Webb immediately ordered Seamans and Phillips fly to the Cape to investigate.

Over five years, nineteen Americans had flown sixteen missions without loss of life – to now lose the first three-man crew in a spacecraft accident on the ground was truly shocking. As Webb observed to newsmen immediately after the fire: "Although everyone realised that some day space pilots would die, who would have thought the first tragedy would be on the ground?"

The Board of Inquiry was headed by Floyd Thompson, the Director of NASA's Langley Research Center, with Borman serving as the Astronaut Office's representative. The general site of the fire's origin was near the foot of Grissom's couch, where components of the ECS had repeatedly been removed and replaced during testing. Although the investigation did not identify the specific ignition source, it did discover physical evidence of electrical arcing in a wiring harness. Evidently, some time during either manufacturing or subsequent testing, an unnoticed incidental contact had scraped the insulation off a wire, providing an opportunity for a spark. The spark had ignited nearby flammable material, and in the super-pressurised oxygen atmosphere the result had been a brief but intense 'flash' fire. There had been some 70 pounds of nylon netting, polyurethane foam and velcro in the capsule, strung around for convenient application, all of it flammable in such an environment. In retrospect, the worst flaw was the inward-opening hatch, which even under ideal conditions took several minutes to open, and would have been *impossible* to open with the cabin pressurised above ambient. Because neither the launch vehicle nor the spacecraft had been loaded with propellants, the 'plugs out' test had not been considered hazardous. Nevertheless, the solid rocket motors in the launch escape system were directly over the spacecraft, and if the heat from the fire had ignited these motors the White Room crew would likely have been killed as well.

On Tuesday, 31 January, Grissom and Chaffee were buried with full military honours at Arlington National Cemetery, and White was buried at West Point.

In an interview for the Associated Press in December 1966, Grissom had told Howard Benedict: "If we die, we want people to accept it. We are in a risky business and we hope that if anything happens to us it won't delay the Program. The conquest of space is worth the risk of life."

In the management reshuffle, Everett Christensen resigned as Apollo Mission Director in Washington D.C. and George Low replaced Joe Shea as Apollo Spacecraft Program Manager. North American Aviation fired Harrison Storms and hired William Bergman from the Martin Company to manage the production of the Apollo spacecraft. The Block I was scrapped and all effort switched to redesigning the Block II, most notably by the incorporation of a hatch that could be swung outward and be opened within seconds.

Reflecting long afterwards, Kraft observed, "There was enough wrong with the [Block I] spacecraft that without the fire we might not have made Kennedy's deadline at all. We would have flown, found some problems, taken months to fix them, flown again, found some more problems, taken more months... We might not have landed on the moon until 1970-71."

Recovery

In reconsidering the crewing, Slayton revised the terminology of the assignments – the Senior Pilot became the Command Module Pilot (CMP) and the Pilot became the Lunar Module Pilot (LMP). At the end of April 1967, Slayton gave Schirra command of the first test of the Block II CSM on a mission lasting up to 14 days – in effect, Grissom's mission – with Stafford backing him up. McDivitt retained the CSM+LM test, now riding with their LM on a Saturn V, the development of which would continue while the spacecraft was redesigned. This mission would be backed up by Pete Conrad, with Dick Gordon as CMP and C.C. Williams as LMP.[10] A high-apogee mission designed to simulate a lunar return was to be flown by a newly formed crew commanded by Borman, with Collins as CSM and Bill Anders as LMP, backed up by another new crew of Neil Armstrong, Jim Lovell as CMP and Buzz Aldrin as LMP.

The first Saturn V thundered off Pad 39 on 9 November 1967 as Apollo 4, and following a high apogee CSM 017 was boosted back into the atmosphere to simulate a lunar return and thereby test the heatshield. This launch vindicated Mueller's 1963 decision to utilise 'all up' testing in which all the components were flight-worthy vehicles. Phillips said at a press conference immediately afterwards that Apollo was now "on its way to the Moon". On 22 January 1968 LM-1 was sent up to test its propulsion system. To fit the LM on the Saturn IB, it had to be flown without its legs. Although several problems were encountered, they were overcome and the results prompted Mueller to cancel the second such mission. A second Saturn V test on 4 April also suffered some problems, but von Braun's team readily identified the causes and fixed them.

In May, Stafford was assigned the first mission expected to fly into cislunar space, and Gordon Cooper was given command of the backup crew, with Donn Eisele as CMP and Ed Mitchell as LMP.[11] In July, when it became clear that McDivitt's CSM+LM test would not be able to be flown before the turn of the year, George Low initiated a study of how a CSM could be sent to orbit around the Moon in December to pre-empt a circumlunar mission that the Soviets were preparing for. In August, McDivitt was offered this alternative mission but rejected it, preferring to wait for the first LM; Borman accepted it. By this time Collins had been withdrawn from Borman's crew to have surgery to his shoulder, Lovell had moved up, and Fred Haise had been assigned in *his* place in backup. Apollo 7 was set for 22 October. A fortnight before this, Schirra announced his retirement from NASA and the Navy, effective upon his return. After Schirra called Apollo 7 a "101 per cent success", Webb agreed to the plan to send Apollo 8 to orbit the Moon in December. However, due to the

---

[10] Unfortunately, Williams was killed in an aircraft accident on 6 October 1967 and the LMP slot on Conrad's crew was assigned to Al Bean, who was serving as the Astronaut Office's representative in the newly formed Apollo Applications Program, which was planning a wide variety of possible missions to follow on from the first lunar landing.

[11] Although Eisele was still a rookie at the time of this assignment, he would be rotating from Schirra's crew. Nevertheless, because he would still have no experience of rendezvous it seems highly unlikely that Slayton would have permitted this backup crew to step forward if something were to ground such an experienced prime crew; so Cooper's crew seemed unlikely to remain together for long.

arcana of orbital dynamics, the launch window for a mission from Kazakhstan opened a fortnight earlier than that for Florida. NASA held its breath and waited, and was delighted when the Soviet's did nothing. As Apollo 8 orbited the Moon on Christmas Eve, its crew read the opening verses from the *Book of Genesis*. In the first week of January 1969, Slayton rotated Armstrong's crew (with Collins, now recovered from his injury, being reinstated over Haise) to Apollo 11. When the LM separated from its CSM on Apollo 9 in March, it marked the first time that a crew flew a spacecraft which was incapable of returning to Earth – if a rendezvous proved impossible, McDivitt and Schweickart would be doomed; however, the Gemini experience paid off. In May, Apollo 10 flew in lunar orbit to rehearse everything to the point of starting the powered descent.[12] Then on 16 July, with less than six months remaining to the decade, Armstrong, Collins and Aldrin set off in Apollo 11 to attempt the first lunar landing.

Charles Mathews, Wernher von Braun, George Mueller and Samuel Phillips celebrate as Apollo 11 sets off for the Moon on 16 July 1969 (left). The result (right) a boot imprint in the lunar dust on 21 July. Without the experience gained during the Gemini Program it is questionable whether Kennedy's challenge would have been met.

Final thought

If NASA had managed to launch Al Shepard a few weeks before the Soviet Union flew Yuri Gagarin, Kennedy may well not have challenged America to land a man on the Moon before the decade was out. The fact that Shepard's flight had been only suborbital would probably not have mattered, as the world's first 'spaceman' would have been an American. The irony is that Shepard could have been sent up a month or so earlier, but an extra test had been scheduled to demonstrate the reliability of the Mercury–Redstone combination. The early months of 1961 therefore serve to illustrate that history is not an irresistible tide, it is extremely sensitive to the outcome of singular events.

---

[12] Because the software for the powered descent was not ready, Stafford could not have landed even if he had wanted to.

Gemini chronology

Spacecraft	Launched	Crew	Duration	Revs
Gemini 1	8 April 1964	-	-	3
Gemini 2	19 January 1965	-	18m 16s	0
Gemini 3	23 March 1965	Gus Grissom John Young	4h 53m	3
Gemini 4	3 June 1965	Jim McDivitt Ed White	97h 56m	62
Gemini 5	21 August 1965	Gordon Cooper Pete Conrad	190h 56m	120
Gemini 7	4 December 1965	Frank Borman Jim Lovell	330h 35m	206
Gemini 6	15 December 1965	Wally Schirra Tom Stafford	25h 51m	16
Gemini 8	16 March 1966	Neil Armstrong Dave Scott	10h 42m	7
Gemini 9	3 June 1966	Tom Stafford Gene Cernan	72h 21m	45
Gemini 10	18 July 1966	John Young Mike Collins	70h 47m	43
Gemini 11	12 September 1966	Pete Conrad Dick Gordon	71h 17m	44
Gemini 12	11 November 1966	Jim Lovell Buzz Aldrin	94h 35m	59

Extravehicular Activity

Crewman	Mission	Duration
Alexei Leonov	Voskhod 2	10m
Ed White	Gemini 4	36m
Gene Cernan	Gemini 9	2h 7m
Mike Collins	Gemini 10	1h 30m (=50m +40m)
Dick Gordon	Gemini 11	2h 41m (=33m +2h 8m)
Buzz Aldrin	Gemini 12	5h 30m (=2h 27m +2h 8m +55m)

Glossary

ACS	Attitude Control System (Agena)
AMU	Astronaut Manoeuvring Unit
ATDA	Augmented Target Docking Adapter
BEF	Blunt End Forward (Gemini)
CC	Corrective Combination (manoeuvre)
CSM	Command and Service Modules (Apollo)
DCS	Digital Command System (Gemini)
ECS	Environmental Control System (Gemini)
ELSS	EVA Life Support System
ESP	EVA Support Package
ETR	Eastern Test Range
EVA	ExtraVehicular Activity
GATV	Gemini-Agena Target Vehicle
GLV	Gemini Launch Vehicle
G-T	Gemini-Titan
HHMU	Hand-Held Manoeuvring Unit
HF	High Frequency (radio)
IMU	Inertial Measurement Unit (Gemini)
IRFNA	Inhibited Red Fuming Nitric Acid (oxidiser)
IVAR	Insertion Velocity Adjustment Routine (manoeuvre)
IVI	Incremental Velocity Indicator (Gemini)
KSC	Kennedy Space Center
LM	Lunar Module (Apollo)
LOR	Lunar Orbit Rendezvous
LOX	Liquid oxygen
MAP	Message Acceptance Pulse (indicator)
MCC	Mission Control Center
MDIU	Manual Data Insertion Unit (Gemini)
MDS	Malfunction Detection System (Titan)
MOCR	Mission Operations Control Room
MSC	Manned Spacecraft Center
NSR	Normal Slow Rate (manoeuvre)
OAMS	Orbital Attitude and Manoeuvring System
PPS	Primary Propulsion System (Agena)
PQI	Propellant Quantity Indicator (Gemini)
RCS	Re-entry Control System (Gemini)
RP1	Rocket propellant No.1
SDP	Status Display Panel (Agena)
SECO	Sustainer Engine Cut-Off (Atlas)
SECO	Second-stage Engine Cut-Off (Titan)
SEF	Small End Forward (Gemini)
SPC	Stored Program Controller (Agena)
SPS	Secondary Propulsion System (Agena)
TDA	Target Docking Adapter

TIROS	Television InfraRed Operational System (meorological satellites)
TPF	Terminal Phase Finalisation
TPI	Terminal Phase Initiation
UDMH	Unsymmetrical dimethyl hydrazine (fuel)
UHF	Ultra High Frequency (radio)
VCM	Ventilation Control Module
WTR	Western Test Range
WWTN	World-Wide Tracking Network

Reading List

The following books and magazine articles (in chronological order) are well worth seeking out.

Into orbit, The Mercury Seven, Cassell, 1962

Project Gemini, Irwin Stambler, Putnam, 1964

Gemini Summary Conference, NASA-SP-138, NASA, 1967

Gemini, Virgil 'Gus' Grissom, Macmillan, 1968

Soviet space exploration - The first decade, William Shelton, Arthur Barker, 1969

The Russian space bluff, Leonid Vladimirov, Tom Stacey, 1971

Soviets in Space - The story of the Salyut and the Soviet approach to present and future space travel, Peter Smolders, Lutterworth Press, 1973

Carrying the fire - An astronaut's autobiography, Michael Collins, W.H. Allen, 1975

On the Shoulders of Titans - A history of Project Gemini, Barton C. Hacker and James M. Grimwood, NASA-SP-4203, NASA, 1977

The history of manned space flight, David Baker, New Cavendish Books, 1981

Liftoff - The story of America's adventure in space, Michael Collins, Grove Press, 1988

Men from Earth, Buzz Aldrin and Malcolm McConnell, Bantam, 1989

DEKE! , Donald K. Slayton with Michael Cassutt, Forge Paperback, 1994

Korolev, James Harford, John Wiley, 1997

Flying the Gusmobile, D.C. Agle, Air & Space, August/September 1998

The Race - The definitive story of America's battle to beat Russia to the Moon, James Schefter, Century, 1999

The last man on the Moon, Eugene Cernan and Don Davis, St Martin's Press, 1999

Challenge to Apollo - The Soviet Union and the Space Race: 1945-1974, Asif Siddiqi, SP-2000-4408, NASA, 2000

A brief history of the Atlas rocket vehicle, Richard Martin, Quest, vol. 8, no. 2, p. 54, 2000

Titan II - A history of a Cold War missile program, David K. Stumpf, University of Arkansas Press, 2000

Project Mercury, John Catchpole, Springer-Praxis, 2001

Gemini - Steps to the Moon, David J. Shayler, Springer-Praxis, 2001

Failure is not an option, Gene Kranz, Simon & Schuster, 2001

Flight - My life in Mission Control, Chris Kraft, Penguin, 2002

We have capture - Tom Stafford and the Space Race, Tom Stafford with Michael Cassutt, Smithsonian Institution Press, 2002

The moonlandings - An eyewitness account, Reginald Turnill, Cambridge University Press, 2003

Apogee Books Space Series

#	Title	ISBN	Bonus	US$	UK£	CN$	
1	Apollo 8	1-896522-66-1	CDROM	$18.95	£13.95	$25.95	_____
2	Apollo 9	1-896522-51-3	CDROM	$16.95	£12.95	$22.95	_____
3	Friendship 7	1-896522-60-2	CDROM	$18.95	£13.95	$25.95	_____
4	Apollo 10	1-896522-52-1	CDROM	$18.95	£13.95	$25.95	_____
5	Apollo 11 Vol 1	1-896522-53-X	CDROM	$18.95	£13.95	$25.95	_____
6	Apollo 11 Vol 2	1-896522-49-1	CDROM	$15.95	£10.95	$20.95	_____
7	Apollo 12	1-896522-54-8	CDROM	$18.95	£13.95	$25.95	_____
8	Gemini 6	1-896522-61-0	CDROM	$18.95	£13.95	$25.95	_____
9	Apollo 13	1-896522-55-6	CDROM	$18.95	£13.95	$25.95	_____
10	Mars	1-896522-62-9	CDROM	$23.95	£18.95	$31.95	_____
11	Apollo 7	1-896522-64-5	CDROM	$18.95	£13.95	$25.95	_____
12	High Frontier	1-896522-67-X	CDROM	$21.95	£17.95	$28.95	_____
13	X-15	1-896522-65-3	CDROM	$23.95	£18.95	$31.95	_____
14	Apollo 14	1-896522-56-4	CDROM	$18.95	£15.95	$25.95	_____
15	Freedom 7	1-896522-80-7	CDROM	$18.95	£15.95	$25.95	_____
16	Space Shuttle STS 1-5	1-896522-69-6	CDROM	$23.95	£18.95	$31.95	_____
17	Rocket Corp. Energia	1-896522-81-5		$21.95	£16.95	$28.95	_____
18	Apollo 15 - Vol 1	1-896522-57-2	CDROM	$19.95	£15.95	$27.95	_____
19	Arrows To The Moon	1-896522-83-1		$21.95	£17.95	$28.95	_____
20	The Unbroken Chain	1-896522-84-X	CDROM	$29.95	£24.95	$39.95	_____
21	Gemini 7	1-896522-80-7	CDROM	$19.95	£15.95	$26.95	_____
22	Apollo 11 Vol 3	1-896522-85-8	DVD*	$27.95	£19.95	$37.95	_____
23	Apollo 16 Vol 1	1-896522-58-0	CDROM	$19.95	£15.95	$27.95	_____
24	Creating Space	1-896522-86-6		$30.95	£24.95	$39.95	_____
25	Women Astronauts	1-896522-87-4	CDROM	$23.95	£18.95	$31.95	_____
26	On To Mars	1-896522-90-4	CDROM	$21.95	£16.95	$29.95	_____
27	Conquest of Space	1-896522-92-0		$23.95	£19.95	$32.95	_____
28	Lost Spacecraft	1-896522-88-2		$30.95	£24.95	$39.95	_____
29	Apollo 17 Vol 1	1-896522-59-9	CDROM	$19.95	£15.95	$27.95	_____
30	Virtual Apollo	1-896522-94-7		$19.95	£14.95	$26.95	_____
31	Apollo EECOM	1-896522-96-3		$29.95	£23.95	$37.95	_____
32	Visions of Future Space	1-896522-93-9	CDROM	$27.95	£21.95	$35.95	_____
33	Space Trivia	1-896522-98-X		$19.95	£14.95	$26.95	_____
34	Interstellar Spacecraft	1-896522-99-8		$24.95	£18.95	$30.95	_____
35	Dyna-Soar	1-896522-95-5	DVD*	$32.95	£23.95	$42.95	_____
36	The Rocket Team	1-894959-00-0	DVD*	$34.95	£24.95	$44.95	_____
37	Sigma 7	1-894959-01-9	CDROM	$19.95	£15.95	$27.95	_____
38	Women Of Space	1-894959-03-5	CDROM	$22.95	£17.95	$30.95	_____
39	Columbia Accident Rpt	1-894959-06-X	CDROM	$25.95	£19.95	$33.95	_____
40	Gemini 12	1-894959-04-3	CDROM	$19.95	£15.95	$27.95	_____
41	The Simple Universe	1-894959-11-6		$21.95	£16.95	$29.95	_____
42	New Moon Rising	1-894959-12-4	DVD*	$33.95	£23.95	$44.95	_____
43	Moonrush	1-894959-10-8		$24.95	£17.95	$30.95	_____
44	Mars Volume 2	1-894959-05-1	DVD*	$28.95	£20.95	$38.95	_____
45	Rocket Science	1-894959-09-4		$TBA	£TBA	$TBA	_____
46	How NASA Learned	1-894959-07-8		$25.95	£18.95	$35.95	_____
47	Virtual LM	1-894959-14-0	CDROM	$TBA	£TBA	$TBA	_____
48	Deep Space	1-894959-15-9	DVD*	$TBA	£TBA	$TBA	_____

CG Publishing Inc home of **Apogee Books**
P.O Box 62034 Burlington, Ontario L7R 4K2, Canada
TEL. 1 905 637 5737   FAX 1 905 637 2631
e-mail marketing@cgpublishing.com
* NTSC Region 0

Many more to come! Check our website for new titles.
www.apogeebooks.com

Check us out on the Web -
http://www.apogeebooks.com

Return this completed form and become eligible to win free books!

One new space book almost every month! The world's number one space book publisher!

Check us out on the Web -
http://www.apogeebooks.com

Return this completed form and become eligible to win free books!

One new space book almost every month! The world's number one space book publisher!

BUSINESS REPLY MAIL

FIRST CLASS MAIL   PERMIT NO. 350 WHEATON IL

POSTAGE WILL BE PAID BY ADDRESSEE

COLLECTOR'S GUIDE PUBLISHING INC
P.O. BOX 4588
WHEATON, IL 60189-9937

NO POSTAGE
NECESSARY
IF MAILED
IN THE
UNITED STATES

THE SPACE BOOK COMPANY!

P.O. BOX 4588
WHEATON ILLINOIS 60189-9937
HTTP://WWW.APOGEEBOOKS.COM

Apogee Books is THE space book company, with almost one NEW space related book published every month.

☐ *To receive a catalog by mail check here and print your address below.*

☐ *Please update me by email (Print email address below)**

NAME: (Please PRINT)

Company:

Address:

City: **State:**   **ZIP:**

E Mail:

*CG Publishing will not disburse this email address to any other organizations for unrelated purposes except as marked below.

Are you are interested in hearing from any Space Advocacy Groups? Check the boxes below.

Space Frontier Foundation ☐   Planetary Society ☐   National Space Society ☐   All available ☐

1

THE SPACE BOOK COMPANY!

P.O. BOX 4588
WHEATON ILLINOIS 60189-9937
HTTP://WWW.APOGEEBOOKS.COM

Apogee Books is THE space book company, with almost one NEW space related book published every month.

☐ *To receive a catalog by mail check here and print your address below.*

☐ *Please update me by email (Print email address below)**

NAME: (Please PRINT)

Company:

Address:

City: **State:**   **ZIP:**

E Mail:

*CG Publishing will not disburse this email address to any other organizations for unrelated purposes except as marked below.

Are you are interested in hearing from any Space Advocacy Groups? Check the boxes below.

Space Frontier Foundation ☐   Planetary Society ☐   National Space Society ☐   All available ☐

2